SpringerBriefs in Applied Sciences and Technology

Thermal Engineering and Applied Science

Series Editor

Francis A. Kulacki

For further volumes:
http://www.springer.com/series/10305

Efstathios E. (Stathis) Michaelides

Heat and Mass Transfer in Particulate Suspensions

 Springer

Efstathios E. (Stathis) Michaelides
Department of Engineering
Texas Christian University
Fort Worth, TX, USA

ISSN 2191-530X   ISSN 2191-5318 (electronic)
ISBN 978-1-4614-5853-1 ISBN 978-1-4614-5854-8 (eBook)
DOI 10.1007/978-1-4614-5854-8
Springer New York Heidelberg Dordrecht London

Library of Congress Control Number: 2012953259

Printed on acid-free paper

Springer is part of Springer Science+Business Media (www.springer.com)

To my wife, Laura

Preface

The subject of heat and mass transfer in particulate suspensions is one of practical as well as intellectual curiosity. The operation of several traditional, large-scale engineering systems, such as industrial dryers and fluidized bed combustors, depends to a high degree on the flow, heat, and mass transfer between particles and fluid and between the fluidized suspension and its boundaries. The world economy depends daily on catalytic crackers in all refineries worldwide for the supply of gasoline and other petroleum products. Also the detailed knowledge and the making of deterministic predictions on the flow, heat, and mass transfer of particle suspensions in gases and liquids is a subject that has intrigued the scientists and engineers for a long time, starting with the works of Fourier, Poisson, Green, and Stokes. At the beginning of the twenty-first century, the scientific and engineering interest on this subject continues unabated and has reached particles of nano-sizes. The latest interest is driven by the desires to optimize the traditional systems and to design particulate systems for new applications, such as heat transfer with nanofluids, which may solve the cooling problem of the next generation microchips and the directed drug delivery to specific organs of humans and animals that will make medicine more effective.

This short monograph is composed of four chapters, the first two on significant theoretical and experimental results of particulate suspensions and the last two on engineering applications. The first chapter includes the most significant analytical and experimental results on the flow, heat, and mass transfer of suspensions. The material presented covers the timescales and pertinent dimensionless parameters, the implications of thermodynamic equilibrium for droplet suspensions, the steady and transient processes, the creeping flow and inertial/advection processes, and the processes with and without mass exchange. The second chapter is an exposition of the numerical methods and tools we employ in multiphase research and modeling. The chapter commences with the desired attributes of multiphase flows; presents the types of models that are needed to describe dilute, intermediate, and dense suspensions; exposes the reader to the most commonly used models for the accurate description of particulate suspension processes; and explains the several numerical methods we use for the accurate modeling of rigid and deformable boundaries of particles. A good description is offered on the topics of particle collisions and droplet coalescence, which are important in the flow and energy exchange of the suspensions. The third chapter is devoted to the fluidized bed reactors, a class of very important and economically valuable engineering systems used in the chemical production, food production, and power production industries. The various types of fluidized bed reactors are described as well as the fundamental processes of particle–fluid momentum and thermal interactions. Finally, the fourth chapter is devoted to a new application of heat transfer in suspensions, the nanofluids that contain nano-size particles. This topic is the subject of a great deal of ongoing research, and most of the results are new and have not been demonstrated in engineering systems. For this reason, this chapter is written as a survey of the subject. It includes few definitive, quantitative results and focuses on the exposition of several sets of experimental data, numerical results, and analytical studies, some of which contradict each other.

While most of the texts on particulate suspensions primarily describe the flow of the suspensions and the mechanical particle–fluid interactions, the focus of this monograph is on the energy and the mass transfer of the suspensions to external systems as well as the heat interactions between fluids and particles. Sufficient information on the flow and momentum interactions of the suspension is given for the reader to have a better understanding of the heat and mass transfer processes and, especially, of the heat and mass advection processes. The monograph is intended for the use of researchers in multiphase flow, who wish to broaden their knowledge to include the heat and mass transfer processes; for the use of engineers and practitioners who wish to learn about the latest developments on the subject; and for graduate students and researchers in the disciplines of mechanical and chemical engineering, who must be familiar with the latest developments and publications.

A number of individuals have helped in this project: My research students, from whom I have learned more than they have learned from me. I am very thankful to my colleagues at TCU and the University of Texas at San Antonio, especially to Dr. Zhi-Gang Feng, for several fruitful discussions on this subject. Ms. Teresa Berry assisted me a lot with the references of the last chapter. The Tex Moncrief Chair at TCU enabled me to devote sufficient effort to this project and finish the manuscript on time. I am also very indebted to my own family, not only for their constant support, but also for lending a hand whenever it was needed. My wife, Laura, has been a constant source of inspiration and help. My son Dimitri, who decided to become a nuclear engineer, devoted a good part of his vacation time to the manuscript and helped with the references. My son Emmanuel and daughter Eleni were always there and ready to help. I owe to all my sincere gratitude for their contributions to this short monograph.

Forth Worth, TX, USA   Efstathios E. (Stathis) Michaelides

Contents

Chapter 1
Fundamentals

Keywords Length-scales · Time-scales · Creeping flow · Inertia · Drag coefficients · Heat transfer coefficients · Thermometers

1.1 Introduction

Applications of the flow, heat, and mass transfer of particles, bubbles, and drops are omnipresent in everyday life and engineering practice. Diverse natural and engineering systems, ranging from nuclear reactors to internal combustion engines, from petroleum refining to sediment and pollutant transport in aquatic environments, and from pharmaceutical production to nanotechnology, involve carrier fluids that convey dispersed materials of another phase, in the form of particles, bubbles, and drops. The design and optimization of engineering systems and the understanding of the operation of these systems necessitate the knowledge of the fundamental processes that pertain to the flow, mass, and heat transfer from particles, bubbles, and drops.

In a heterogeneous mixture of two or more phases, the constituents have distinct physical properties and, in general, move with different velocities. The constituents of a flowing heterogeneous mixture, such as a particulate suspension, always exchange linear and angular momentum, oftentimes exchange mass, and also exchange energy. The processes of momentum, energy, and mass interactions are always related in multiphase engineering systems. For example, in a direct contact heat exchanger, where colder drops are sprayed in the midst of vapor to be condensed, the drops absorb enthalpy from the vapor, and thus, their temperature increases. Because of the direct contact between the cooler drops and the vapor, some of the mass of the vapor condenses on the surface of the drops, thus, increasing the average size of the drops. And, as a result of the hydrodynamic interaction between the vapor and the drops, or between multiple drops, larger drops may break up in two or more smaller ones.

E.E. (Stathis) Michaelides, *Heat and Mass Transfer*
in Particulate Suspensions, SpringerBriefs in Applied Sciences and Technology,
DOI 10.1007/978-1-4614-5854-8_1, © Springer Science+Business Media New York 2013

While it is possible to derive general equations for the exchange of mass, momentum, and heat in all dispersed multiphase flow applications, because of the complexity of most practical systems, it is difficult, and often impossible, to obtain an exact solution of these equations in the most general cases, without the use of simplifying assumptions that restrict the generality of the solutions. This does not pose a significant problem for the vast majority of systems and applications, because engineers and scientists are not interested in all the details of the flow and the transport processes, but in the specific macroscopic characteristics and properties of the multiphase system, which are needed for the design of the system or for the optimization of the process. For this reason, in the solution of the multiphase flow-governing equations, and specifically in the heat and mass transfer processes involving particles and drops, the engineering interest pertains to particular aspects of the heat and mass transfer processes that answer specific scientific or technical questions.

1.1.1 Nomenclature

A.   **Latin Symbols**
α_s Absorptivity
A Area
B Blowing factor
c Specific heat capacity
C_D Drag coefficient
D Conduit diameter
$d = 2\alpha$ Particle diameter
D_f Mass diffusion coefficient
f Frequency
F Dimensionless drag force
g Gravitational acceleration
h Enthalpy
h_c Convective heat transfer coefficient
h_m Mass transfer coefficient
k Thermal conductivity
L Lengthscale of fluid
m Mass
Nu Nusselt number
P Pressure
Pe Peclet number
Re Reynolds number
t Time

T	Temperature
U, u	Fluid velocity
V, v	Particle velocity
x, y, z	Coordinates
Y	Mass fraction

B.   Greek symbols

α	Radius
ε	Emissivity
ζ	Accommodation coefficient
λ	Viscosity ratio
μ	Dynamic viscosity
ν	Kinematic viscosity
ξ	Interface thermal slip
ρ	Density
σ	Surface tension
τ	Timescale
ϕ	Volume fraction of particles
ψ	Stream function
Ψ	Shape factor
Ω	Angular velocity

C.   Subscripts

e	Effective
f	Fluid
fg	Latent heat/enthalpy
m	Pertains to film properties
M	Pertains to momentum
mf	Minimum fluidization
mol	Molecular
opt	Optimum
rot	Rotational
th	Thermal
s	Sphere/particle
T	Total

D.   Superscripts

0	Undisturbed field
'	Fluctuation
.	Time rate
_	Time-averaged

1.1.2   Timescales, Lengthscales, and Dimensionless Groups

Problems involving particulate flows and heat transfer entail at least two lengthscales: the radius of the particle, α, and the characteristic lengthscale of the fluid, L. Oftentimes there are multiple lengthscales for the fluid as, for example, in turbulent channel flows, where the dimensions of the channel, the height and width, the Kolmogorov scale, and the viscous dissipation scale play important roles in the transport processes of the fluid.

Similarly, there are multiple timescales related to the flow and heat transfer of particulate systems. Of these, the thermal timescale of the fluid and the thermal timescale of the particles are typically the most important in heat transfer processes. Because the advective heat transfer depends on the relative motion of the particles and the fluid, the timescales of the motion of particles and fluid also affect significantly the heat and mass transfer processes between the particles and the fluid. The thermal timescales of the particle and the fluid are defined, respectively:

$$\tau_{\text{th}} = \frac{\alpha^2 \rho_s c_s}{3 k_f} \text{ and } \tau_{\text{fth}} = \frac{4 L^2 \rho_f c_f}{k_f}. \tag{1.1}$$

The timescales for the equation of motion of the particle and the advection timescale for the fluid are, respectively:

$$\tau_M = \frac{2 \alpha^2 \rho_s}{9 \mu_f} \text{ and } \tau_{fM} = \frac{L}{u}. \tag{1.2}$$

It must be noted that the definition of timescales is rather arbitrary and that some authors have adopted definitions with different numerical coefficients. The particulate timescales in the last two expressions were chosen so as to render the coefficients of the drag and conduction terms in the dimensionless equation of motion and energy equation equal to one (Michaelides 2003). If the fluid undergoes an oscillatory motion, the inverse of the frequency of this motion, $1/f$, is the appropriate timescale for the fluid. In addition to the rectilinear motion, which is characterized by the magnitude of the relative velocity, $|u_i - v_i|$, the particle may undergo rotational motion, which is characterized by a rotational velocity, Ω, or by the local shear of the fluid, γ.

Several dimensionless groups are defined as the ratios of the pertinent timescales or the pertinent forces that govern the physical phenomena and processes involving particulate flow and heat or mass transfer. Among these dimensionless groups are the following.

1.1.2.1   Viscosity Effects

Three Reynolds numbers for the particles are defined with respect to the rectilinear velocity, the rotational velocity, and the local shear. In addition, a separate Reynolds number is defined for the fluid. The first three dimensionless groups are based on the particle radius and the last on the characteristic lengthscale of the fluid, L.

$$Re_s = \frac{2\alpha\rho_f|\vec{u}-\vec{v}|}{\mu_f}, \quad Re_\gamma = \frac{4\alpha^2\rho_f\gamma}{\mu_f}, \quad Re_{rot} = \frac{4\alpha^2\rho_f|\vec{\Omega}|}{\mu_f}, \quad Re_L = \frac{L\rho_f|\vec{u}|}{\mu_f}. \quad (1.3)$$

1.1.2.2 Heat and Mass Transfer Effects

The pertinent Peclet number, Nusselt number, Biot number, and Sherwood number for the heat and mass transfer for the particle and the bulk fluid are defined in terms of the diameter of spherical particles, 2α, as follows:

$$Pe_s = \frac{2\alpha\rho_f c_f|\vec{u}-\vec{v}|}{k_f}, \quad Pe_\gamma = \frac{4\alpha^2\gamma\rho_f c_f}{k_f}, \quad Pe_m = \frac{2\alpha|\vec{u}-\vec{v}|}{D_f},$$

$$Pe_f = \frac{L\rho_f c_f|\vec{u}|}{k_f}, \quad Nu = \frac{2\alpha h_c}{k_f}, \quad Bi = \frac{2\alpha h_c}{k_s}, \quad Sh = \frac{2\alpha h_m}{D_f}. \quad (1.4)$$

1.1.2.3 Surface Tension Effects

These are characterized by the Bond number, the capillary number, the Eotvos number, the Morton number, and the Weber number, which are, respectively, defined as follows:

$$Bo = \frac{4g\alpha^2|\rho_s-\rho_f|}{\sigma}, \quad Ca = \frac{\mu_f|\vec{u}-\vec{v}|}{\sigma} = \frac{We}{Re}, \quad Eo = \frac{4\alpha^2 g\rho_f}{\sigma},$$

$$Mo = \frac{g\mu_f^4}{\rho_f\sigma^3}, \quad We = \frac{2\alpha\rho_f|\vec{u}-\vec{v}|^2}{\sigma} = ReCa. \quad (1.5)$$

1.1.2.4 Dimensionless Property Numbers

The Prandtl number, Lewis number, and Schmidt number for the transport properties of the fluid are defined as follows:

$$Pr = \frac{c_f\mu_f}{k_f}, \quad Le = \frac{k_f}{\rho_f c_f D_f}, \quad Sc = \frac{\mu_f}{\rho_f D_f}. \quad (1.6)$$

1.1.2.5 Other Effects

Molecular or rarefaction effects are quantified by the Knudsen number, phase-change effects by the Stefan number, oscillatory effects by the Strouhal number, and particle inertia effects by the Stokes number. These dimensionless groups are defined as follows:

$$Kn = \frac{L_{mol}}{2\alpha}, \quad Ste = \frac{c_s \Delta T}{h_{fg}}, \quad Sl = \frac{2\alpha f}{|\vec{u} - \vec{v}|}, \quad St = \frac{2\alpha \rho_s |\vec{u} - \vec{v}|}{9\mu_f}. \tag{1.7}$$

1.2   Thermodynamics of Phase Change

It is well known that the saturation pressure and temperature of two phases in thermodynamic equilibrium separated by a plane interface are related by a functional relationship $P_{sat} = f(T_{sat})$. The functional relationship between the two is rather complex and, for this reason, it is frequently given in tabular form, in the so-called thermodynamic tables. However, and because of the surface tension, this functional relationship is altered when the interface has a curvature, as, for example, in the case of drops. This section will examine the modification of the classical thermodynamic saturation relationship due to the liquid surface curvature and to the presence of a carrier gas component in the vapor phase.

Let us consider a liquid sphere of radius α, inside a carrier gas of density ρ_f, in a spherical system of coordinates whose center coincides with the center of the sphere. The carrier gas and the sphere may have different chemical compositions, and their species will be denoted by the subscripts c and s, respectively. Mass is allowed to cross the boundary of the sphere due to one of the processes of sublimation, evaporation, or condensation. Because of this, the gaseous phase is composed of both the species s and c, while the sphere, which may be either solid or liquid, is composed solely of the species s. Figure 1.1 is a schematic diagram of such a phase-

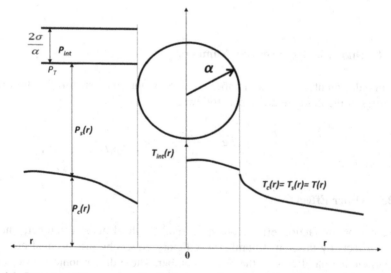

Fig. 1.1 Pressure and temperature distribution around an evaporating drop in a gaseous carrier

change process. The directions and numerical signs of the pressure and temperature gradients in this figure correspond to the process of vapor condensation. The total pressure in the gaseous mixture is constant and equal to P_T. This total pressure is comprised of two parts: the partial pressure, or vapor pressure, of the species s, P_s, and the partial pressure of the carrier gaseous species c, P_c. In all the transport processes that are pertinent to mass transfer to or from the sphere, both partial pressures are functions of the radial direction, r. The total pressure outside the sphere is:

$$P_T = P_c(r) + P_s(r), \tag{1.8}$$

where the parentheses denote arguments of functions, not multiplication of variables. If the carrier phase is composed of a number of several chemical components, the partial pressure, P_c, would be the sum of the partial pressures of all the components. When the sphere is composed of a liquid, we consider it a drop. The pressure in the interior of the drop is augmented by the contribution of the surface tension:

$$P_{int} = P_T + \frac{2\sigma}{\alpha}. \tag{1.9}$$

While in most practical cases the radius of the sphere is large enough to satisfy the condition $P_{int} \approx P_T$, simple calculations show that the interior pressure of drops in the micro- and nano-size range deviates significantly from the total exterior pressure. For example, at 100 °C, the interior pressure of nano-droplets of water with radius $\alpha = 200$ nm would be approximately 0.76 MPa or 7.5 times higher than the saturation pressure pertaining to a flat interface.

With the exception of very fast combustion processes, during most of the engineering applications involving drops and particles, the thermodynamic relaxation time is much shorter than the thermal and momentum characteristic times of the sphere, defined in Eqs. (1.1) and (1.2). For this reason, thermodynamic equilibrium is established fast at the interface of the sphere and the carrier fluid. Thermodynamic equilibrium within the gaseous carrier phase implies that the temperatures of the species s and c are equal, that is, $T_c(r) = T_s(r) = T(r)$, for $r > \alpha$. The temperature inside the sphere in general is nonuniform, e.g., because of heat conduction or heat generation, and hence, it is a function of the radial distance as shown in Fig. 1.1. This temperature will be denoted by the function $T_{int}(r)$, where $r < \alpha$.

At any interface where a change of phase of a species occurs, there is a phase transition, liquid–vapor or solid–vapor, which normally occurs within a very short layer with size of a few molecular lengths. The phase of the matter in this film layer is not well-defined and is often referred to as the "mushy region." The material and thermodynamic properties within this thin layer are not well-defined and are often approximated as linear combinations of the vapor and liquid properties. It is known that the specific properties of matter exhibit a jump discontinuity within this layer, e.g., from liquid enthalpy to vapor enthalpy. The pressure discontinuity is described by Eq. (1.9). Any temperature discontinuity vanishes when thermodynamic equilibrium has been established. Even under thermodynamic nonequilibrium, the temperature difference is very small for larger spheres with $Kn < 0.01$. For this reason, it is often assumed in continuum theory that $T_{int}(\alpha) = T_c(\alpha)$. With very small spheres, in the range $0.01 < Kn < 0.2$, a temperature discontinuity at the interface has been observed experimentally and analytically. Models for the rate of heat transfer use analytical or empirical closure equations to describe this discontinuity (Brunn 1982; Feng and Michaelides 2012). When the size of the sphere is extremely small so that $0.2 < Kn$, the continuum assumptions are not valid and one has to reexamine what the concept of temperature means. The subject of what is a continuum and the characteristics of a continuum are exposed in more detail in Sect. 4.2.

1.2.1 Effect of the Carrier Gas Concentration

Let us consider the interface of a liquid and its own vapor. The thermodynamic equilibrium between the two phases leads to the following condition of the specific Gibbs free energy for the two phases:

$$dg_v = dg_s \Rightarrow -s_v dT + \frac{dP}{\rho_v} = -s_s dT + \frac{dP}{\rho_s}. \qquad (1.10)$$

The last expression yields the so-called *Clausius–Clapeyron equation*, which defines the relationship between the saturation temperature and the saturation pressure at the interface of a flat surface:

$$\frac{dP}{dT}\bigg|_{sat} = \frac{h_{fg}}{T\left[\frac{1}{\rho_v} - \frac{1}{\rho_s}\right]}, \qquad (1.11)$$

where the derivative, dP/dT, is evaluated at the interface; T is the temperature at the interface; ρ_v is the density of the species s in its vapor phase, evaluated at the interfacial pressure, P, and temperature, T; and ρ_s is the density of the liquid species s. The relationship between the saturation pressure and temperature for several common liquids and vapors may be found in thermodynamic tables, which appear as appendices to most textbooks on the subject of Thermodynamics.

Now let us consider the isothermal ($dT = 0$) introduction of the carrier fluid in this phase-change system. This introduction increases the total pressure on the liquid phase from P_{sat} to P_T. The pressure of the vapor increases by a different amount and from P_{sat} to P_s. Since the temperature of the system remains constant at T_{sat}, Eq. (1.10) will yield the following condition for the pressure:

$$dg_v = dg_s \Rightarrow \frac{dP}{\rho_v} = \frac{dP_v}{\rho_s}. \qquad (1.12)$$

If the system is far from the critical point, $\rho_v \ll \rho_s$ and the density of the vapor may be approximated by the ideal gas equation[1] of state ($\rho_v \approx P_v/R_vT$), the last expression yields the following differential equation:

$$\frac{dP}{R_vT\rho_s} = \frac{dP_v}{P_v}.$$
(1.13)

Since the density of the liquid is not a strong function of pressure, the denominator of the first fraction may be considered constant. The last equation may then be integrated within the limits P_{sat} to P_T for the liquid and P_{sat} to P_v for the vapor to yield

$$\ln\frac{P_v}{P_{sat}} = \frac{(P_T - P_{sat})}{R_vT\rho_s} \quad or \quad (P_v - P_{sat}) \approx \frac{\rho_v}{\rho_s}(P_T - P_{sat}).$$
(1.14)

For most of the common fluids, $\rho_v/\rho_s \ll 1$ and the effect of the carrier gas presence on the vapor pressure is not significant. Thus, we may assume in most common cases that, even in the presence of a gas, $P_v = P_s = P_{sat}$.

1.2.2   Effect of Curvature and Surface Tension

As it becomes apparent from Eq. (1.9), the surface tension of the sphere affects significantly the saturation pressure of very small spheres. Thermodynamic equilibrium at the surface of the sphere requires that the liquid of the droplet and the carrier fluid have the same temperature. In this case, the vapor pressure of the species s, P_s, would be higher than the saturation pressure, defined by the flat interface $P_s > P_{sat}$. As in the previous section, an analytical expression for the new saturation vapor pressure may be obtained if the state of the droplet is far from the critical state, which implies that $\rho_v \ll \rho_s$ and, also, $\rho_v \approx P/R_vT$. In this case we may derive the following analytical expression:

$$P_s = P_{sat} \exp\left[\frac{2\sigma}{a\rho_sR_vT}\right] \approx P_{sat}\left[1 + \frac{2\sigma}{a\rho_sR_vT}\right].$$
(1.15)

This correction on the vapor pressure is typically very small for materials with low to medium surface tension, unless the radius of the sphere is in the submicron range. For example, for evaporating water droplets at ambient conditions ($\rho_s = 1,000$ kg/m^3, $T_{sat} = 373$ K, $R_v = 0.462$ kJ/kgK, $h_{fg} = 2,200$ kJ/kg, $\sigma = 0.072$ N/m), a 1% or more increase of the vapor pressure would occur when

---

[1] The use of the ideal gas equation is convenient but not critical in this derivation. Other equations of state or tubular data may be used instead. In the latter case, the final result may only be obtained by numerical integration.

the droplets have radii less than 84 μm. The situation is different for liquid metal droplets with high surface tension and moderate boiling points. For example, evaporating mercury droplets at $P_T = 1$ atm ($\rho_s = 13{,}600$ kg/m^3, $T_{sat} = 630$ K, $R_v = 0.04157$ kJ/kgK, $h_{fg} = 301$ kJ/kg, $\sigma = 0.417$ N/m) would show vapor pressure increases higher than 1% when their radii are less than or equal to 234 μm. Actually, 50 μm droplets of evaporating mercury show a 5% increase of their vapor pressure.

All calculations show that, in the case of nano-droplets, the correction of the vapor pressure due to the curvature is always significant and must be taken into account in computations. For example, the vapor pressure of a water droplet with a radius of 2 μm would increase by 42% and would double for a droplet of radius 0.84 μm. The vapor pressure correction due to the curvature has significant implications on the modeling of the evaporating process of droplets, whose size diminishes. Of course, when the size of the droplets becomes low enough for the Knudsen number to be higher than 0.2, the droplets may not be considered as continua, and the thermodynamic equilibrium assumption is not valid. The modeling of the last stages of the evaporation of droplets may be achieved only by molecular dynamics.

1.3   Equation of Motion

The continuity and momentum equations for any motion inside a viscous fluid are

$$\frac{\partial \rho}{\partial t} + \rho \vec{\nabla} \cdot \vec{u} = 0, \tag{1.16}$$

and

$$-\vec{\nabla} P + \mu \nabla^2 \vec{u} = \rho \left(\frac{\partial \vec{u}}{\partial t} + \vec{u} \cdot \vec{\nabla} \vec{u} \right). \tag{1.17}$$

When immersed objects—particles, bubbles, and drops—are carried by viscous fluids, the two equations may be solved with the appropriate boundary conditions at the interface to determine the flow fields around the immersed objects. The hydrodynamic force acting on the immersed objects and drag coefficients are usually derived from this flow field.

The last equation (1.17) may be rendered dimensionless by using the momentum timescale for the sphere: $\tau_M = 4\alpha^2 \rho_f / \mu_f$. The following dimensionless equation is then derived:

$$-\vec{\nabla}^* P^* + \nabla^{*2} \vec{u}^* = \frac{\partial \vec{u}^*}{\partial t^*} + Re_s \left(\vec{u}^* \cdot \vec{\nabla}^* \vec{u}^* \right), \tag{1.18}$$

where the asterisk denotes a dimensionless variable. Because the size of particles, bubbles, and drops is very small, the condition $Re_s \ll 1$ applies in several particulate systems. In such cases, the last term, which is often called the advection term, may be neglected, and one obtains the following equation, which is referred to as the *Stokes equation*:

$$-\vec{\nabla}^* P^* + \nabla^{*2}\vec{u}^* = \frac{\partial \vec{u}^*}{\partial t^*}. \tag{1.19}$$

This is a linear equation that may be solved analytically to derive results for a type of flow that is referred to as "Stokes flow" or "creeping flow." The term implies that the Reynolds number of the sphere, Re_s, is very small and all the inertia effects have been neglected.

1.3.1   Steady, Stokesian Flow for Spheres

Consider the motion of a small, viscous, fluid sphere inside a fluid of different viscosity that moves slowly enough for the condition $Re_s \ll 1$ to be satisfied. The magnitude of the relative fluid velocity between the sphere and the far-away fluid velocity is denoted by U, and the no-slip condition applies on the surface of the sphere. The center of coordinates coincides with the center of the sphere, and the flow domain is much larger than the diameter of the sphere. This problem was solved independently by Hadamard (1911) and Rybczynski (1911) who obtained the following expressions for the stream functions inside and outside the viscous sphere:

$$\psi_i = \frac{Ur^2(\alpha^2 - r^2)\sin^2\theta}{4(\lambda+1)\alpha^2}, \tag{1.20}$$

and

$$\psi_o = \frac{Ur^2\sin^2\theta}{2}\left[1 - \frac{(3\lambda+2)\alpha}{2(\lambda+1)r} + \frac{\lambda\alpha^3}{2(\lambda+1)r^3}\right], \tag{1.21}$$

where λ is the ratio of the dynamic viscosities, μ_s/μ_f. The case of a solid sphere is given at the limit $\lambda \to \infty$, and the case of an inviscid sphere, which is a good approximation of a gas bubble, is at the limit $\lambda \to 0$. For a solid sphere, the fluid velocity disturbance, which is caused by the presence of the sphere, may be determined from the last two equations to yield the following expressions for the velocity components of the carrier fluid outside the sphere at $r > \alpha$:

$$u_r = U\cos\theta\left[1 - \frac{(3\lambda+2)\alpha}{2(\lambda+1)r} + \frac{\lambda\alpha^3}{2(\lambda+1)r^3}\right], \tag{1.22}$$

and,

$$u_\theta = -U \sin \theta \left[1 - \frac{(3\lambda + 2)\alpha}{4(\lambda + 1)r} - \frac{\lambda \alpha^3}{4(\lambda + 1)r^3} \right].$$ (1.23)

It is evident that the velocity field described by the above two equations yields $u = 0$ on the surface of the sphere (the no-slip condition) and $u = U$ far away from the solid sphere (the far-away condition). Integration of the normal and shear stresses on the surface of the fluid sphere that result from the last two equations determines the hydrodynamic or drag force exerted by the fluid on the sphere:

$$\vec{F}_D = 2\pi\alpha\mu_f \frac{3\lambda + 2}{\lambda + 1} \vec{U}.$$ (1.24)

Thus, the drag coefficient for the viscous sphere in Stokes flow is given by the expression:

$$C_D = \frac{2|\vec{F}_D|}{\pi\rho|\vec{U}|^2 \alpha^2} = \frac{8(3\lambda + 2)}{Re_s(\lambda + 1)}.$$ (1.25)

Equation (1.25) yields the so-called *Stokes drag* for a solid sphere, $C_D = 24/Re_s$, and for an inviscid bubble $C_D = 16/Re_s$. These correspond to drag forces that are equal to $F = 6\pi\alpha U \mu_f$ for the solid sphere and $F = 4\pi\alpha U \mu_f$ for the inviscid bubble. The latter expression is sometimes referred to as the "form drag," and the difference of the two expressions, which is equal to $2\pi\alpha U \mu_f$, is referred to as the "friction drag." While several authors make this distinction between the two parts of the hydrodynamic force, it must be noted that the drag force is a single entity that arises from the interactions between the fluid and the sphere, not two different forces.

Spheres settling or rising under gravity at Stokes flow conditions are subjected to the gravity/buoyancy force and the hydrodynamic/drag force. At steady-state, the two opposing forces are equal in magnitude, and the spheres move at constant velocity, which is called the *terminal velocity*. An expression for the terminal velocity of a viscous sphere in Stokesian flow is as follows:

$$v_t = \frac{2}{3} \frac{g\alpha^2(\rho_f - \rho_s)}{\mu_f} \frac{\lambda + 1}{3\lambda + 2}.$$ (1.26)

In industrial applications, bubbles and drops seldom follow the predictions of the Hadamard–Rybczynski analysis, because of the presence of impurities in the fluid, which act as *surfactants*. The surfactants, in general, tend to dampen the internal velocity field, thus increasing the effective viscosity of the internal fluid. In this case the behavior of the viscous spheres is closer to that of a solid sphere. Experimental data suggest that as long as the flow is characterized as Stokes flow ($Re_s \ll 1$), the viscous spheres behave as solid spheres when the Bond number, Bo, is less than 4. At $Bo > 4$, a transition occurs and the spheres follow more closely the Hadamard–Rybczynski analysis. It must also be pointed out that in the Stokes flow regime, all bubbles and drops remain spherical, regardless of the value of Bo.

1.3.2 Steady Flow at Higher Reynolds Numbers

1.3.2.1 Solid Spheres

Under the Stokesian flow conditions there is a fore-and-aft symmetry in the flow field around the sphere. At finite Re_s, there are stronger advective effects, and this symmetry breaks down. Experimental observations have shown that, even at low (but finite) Reynolds numbers, a wake is formed behind the sphere. The wake becomes stronger as Re_s increases, and the inertia of the flow around the sphere overcomes the viscosity effects on the surface of the sphere. Experimental observations (Taneda 1956; Achenbach 1974; Seeley et al. 1975) as well as recent numerical computations, give sufficient evidence that the following flow descriptions, related to the presence and behavior of the wake, may be observed around solid spheres (Michaelides 2006):

1. *Attached flow*, which occurs in the range $0 < Re_s < 20$. The wake is not visible in this range, the flow is almost symmetric, and the entire flow is attached to the sphere.
2. *Steady-state wake*. The onset of the wake formation occurs at approximately $Re_s = 20$. This wake is very small in volume and is attached to the aft of the sphere. As Re_s increases, the wake becomes larger, and its point of attachment on the sphere moves forward.
3. *Unsteady wake*. The onset of instability for the wake occurs in the range $130 < Re_s < 150$. At this range a weak long-period laminar oscillation appears at the tip of the wake, whose amplitude increases with Re_s. Pockets of vorticity begin to be shed from the tip of the sphere and influence the flow field away from the sphere. The unsteady wake regime has been observed in experiments in the range $(130–150) < Re_s < 270$.
4. *High subcritical range*, which occurs in the range $270 < Re_s < 3 \times 10^5$. Vortices are shed regularly from alternate sides of the sphere. At the lower end of this regime, the Strouhal number of the vortices, Sl, is a monotonic function of Re_s and increases from 0.1 at $Re_s = 400$ to approximately 2 at $Re_s = 6{,}000$. At values $Re_s > 6{,}000$, separation occurs at a point that rotates equal around the sphere with frequency equal to the shedding frequency and with Sl equal to 0.125.
5. *Supercritical flow*. The onset of the transition to a turbulent wake occurs at approximately $Re_s = 2 \times 10^5$, and the transition is completed at approximately $Re_s = 3.7 \times 10^5$. Changes in the flow patterns occur that are referred to as "critical transition." Above the value $Re_s = 3.7 \times 10^5$, the vortex separation points begin to move downstream, and fluctuations in the position of the separation point become evident. The detached free shear layer becomes turbulent and attaches to the surface of the sphere. The most evident result of this change is the sharp drop of the drag coefficient from the value of approximately 0.42 to 0.07. The transition to a turbulent boundary layer is sensitive to the intensity of the free-stream turbulence, and may be accelerated by "tripping" the

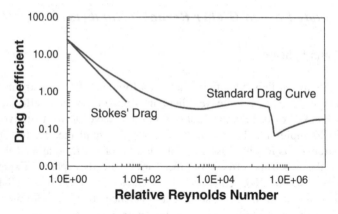

Fig. 1.2 The standard drag curve for a solid sphere

flow, e.g., by a thin wire. This technique has often been used in experimental studies of turbulent boundary layers (Maxworthy 1969).[2]

Regarding the analytical studies on the drag force on spheres at finite Re_s, Proudman and Pearson (1956) employed a singular asymptotic expansion to calculate the velocity field around solid spheres and cylinders at steady state and at $Re_s < 1$. They derived the following expression for the drag coefficient of spheres:

$$C_D = \frac{24}{Re_s}\left[1 + \frac{3}{16}Re_s + \frac{9}{160}Re_s^2 \ln\left(\frac{Re_s}{2}\right) + O\left(Re_s^2\right)\right]. \qquad (1.27)$$

Expressions such as Eq. (1.27) may be used with accuracy for applications in the range $0 < Re_s < 0.7$. At higher values of Re_s, it is advisable to use one of the empirical or semi-empirical expressions for the steady drag coefficient that abound in the literature. One of them is the Schiller and Nauman (1933) correlation, which applies in the range $1 < Re_s < 800$. This correlation is relatively simple and accurate and covers most of the range of particulate flow applications:

$$C_D = \frac{24}{Re_s}\left(1 + 0.15Re_s^{0.687}\right). \qquad (1.28)$$

The standard drag curve for a solid sphere, which is depicted in Fig. 1.2, is a graphical representation of the drag coefficient for smooth, solid spheres. More information on the wakes formed behind the solid and fluid spheres and the several expressions that have been derived for the drag coefficient may be found in Michaelides (2006).

---

[2] It must be noted that all the values of Re_s at the transition points of all the flow regimes are approximate. The values and ranges quoted are for a smooth sphere. The roughness of the surface of the sphere plays an important role in the actual transitions.

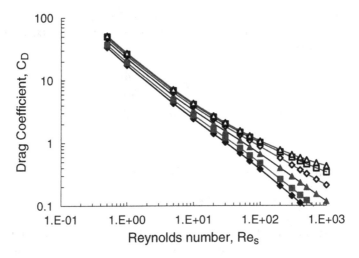

Fig. 1.3 Drag coefficients for viscous spheres for several values of the viscosity ratio, λ. *From bottom to top*: $\lambda = 0, 0.25, 1, 3, 10$ and ∞

1.3.2.2 Fluid, Viscous Spheres

At higher Re_s, the external flow creates an internal circulation, which affects the drag coefficient. The viscosity ratio, $\lambda = \mu_s/\mu_f$, is the primary parameter that defines the internal and external circulation and, plays an important role in the determination of the drag coefficients of viscous spheres, such as bubbles and drops. Feng and Michaelides (2001a) performed a numerical study, using the two-layer concept for the computational grid, and were able to perform accurate computations that extend to $Re_s = 1,000$, beyond which a boundary layer is well formed on the outside surface of the viscous sphere. They used their computational results to derive simple engineering correlations for C_D in terms of λ and Re_s. Figure 1.3 depicts some of these results for several values of the viscosity ratio, λ. The standard drag coefficient curve for solids spheres in this figure corresponds to the limit $\lambda \to \infty$. It is apparent from this figure and the available numerical and experimental data that the drag coefficient of fluid spheres with $\lambda > 10$ is almost equal to the drag coefficient of solid spheres. Therefore, one may model the viscous spheres in low-viscosity carrier fluids as solid spheres. Another observation from this figure is that, at the low end of the Re_s range, the drag coefficient of inviscid bubbles is two-thirds of the drag coefficient of solid spheres, with viscous bubbles and drops being within this narrow range of C_D values. At higher Re_s, the range of the C_D values increases and covers a full decade.

It must be noted that, as it has been experimentally observed, the presence of impurities or surfactants in the carrier fluid makes most interfaces to behave as solid surfaces and that, oftentimes, bubbles and drops behave as solid spheres. The presence of surfactants also affects the shape and the sideways (transverse) motion of bubbles in liquids (Michaelides 2006).

When expressed in terms of the relative Reynolds numbers, the correlations derived by Feng and Michaelides (2001a) are valid up to $Re_s = 1,000$ and may be summarized as follows:

$$C_D(Re_s, \lambda) = \frac{2 - \lambda}{2} C_D(Re_s, 0) + \frac{4\lambda}{6 + \lambda} C_D(Re_s, 2)$$

$$\text{for } 0 \leq \lambda \leq 2, \text{ and } 5 < Re_s \leq 1,000,$$

(1.29)

and

$$C_D(Re_s, \lambda) = \frac{4}{\lambda + 2} C_D(Re_s, 2) + \frac{\lambda - 2}{\lambda + 2} C_D(Re_s, \infty) \quad \text{for}$$

$$2 \leq \lambda \leq \infty, \text{ and } 5 < Re_s \leq 1,000,$$

(1.30)

where the functions $C_D(Re_s, 0)$, $C_D(Re_s, 2)$, and $C_D(Re_s, \infty)$ represent the drag coefficient for inviscid bubbles, the drag coefficient for viscous drops with $\lambda = 2$, and the drag coefficient for solid spheres, respectively. The following functions are recommended to be used with the last two correlations:

$$C_D(Re_s, 0) = \frac{48}{Re_s} \left(1 + \frac{2.21}{\sqrt{Re_s}} - \frac{2.14}{Re_s} \right),$$

(1.31)

$$C_D(Re_s, 2) = 17.0 Re_s^{-2/3},$$

(1.32)

and

$$C_D(Re_s, \infty) = \frac{24}{Re_s} \left(1 + \frac{1}{6} Re_s^{2/3} \right).$$

(1.33)

The expressions for $C_D(Re_s, 0)$ and $C_D(Re_s, \infty)$ above are commonly used correlations for the drag coefficient of bubbles and solid particles. The expression for the drag coefficient at $\lambda = 2$ is a simple correlation of the numerical results. In the low Re_s range, $0 < Re_s < 5$, which is not covered by the above expressions, the following expression is recommended (Feng and Michaelides 2001a):

$$C_D = \frac{8}{Re_s} \frac{3\lambda + 2}{\lambda + 1} \left(1 + 0.05 \frac{3\lambda + 2}{\lambda + 1} Re_s \right) - 0.01 \frac{3\lambda + 2}{\lambda + 1} Re_s \ln(Re_s)$$

$$\text{for } 0 \leq Re_s \leq 5.$$

(1.34)

Equation (1.34) has been derived from the results of the numerical computations in a way that it reduces asymptotically to the Hadamard–Rybczynski solution at $Re_s = 0$ and to the Oliver and Chung (1987) expression for small Re_s. The functional form of the last expression has been derived from the natural next-order asymptotic expansion in terms of Re_s (Proudman and Pearson 1956). Although the last expression was derived from numerical results in the range

$1 < Re_s < 5$, calculations have shown that it accurately predicts the drag coefficient up to $Re_s = 20$ and may be used in the range $Re_s < 20$ instead of expressions (1.29) and (1.30). The maximum fractional difference of the drag coefficient correlations of Eqs. (1.29) through (1.34) is 4.6%, and the standard deviation of all the fractional differences is 2.1%.

It is apparent from these correlations that the density ratio ρ_f/ρ_s does not influence significantly the drag coefficient of viscous spheres. Feng and Michaelides (2001a) determined that, while keeping all the other parameters constant, the variation of the density ratio by two orders of magnitude had an effect of less than 2% on the values of the drag coefficient.

At the higher range of the Reynolds numbers, the wake behind the spheres becomes unsteady and a transient expression for the drag force should be used, if one is known. In such cases the above expressions may be used for the steady-state part of the total hydrodynamic force. The correlations would yield a good approximation for the total hydrodynamic force in transient flow, only when the steady-state part is significantly greater than the other parts or when the timescales of the transients are low enough for the flow to be considered quasi-steady.

It is well known that, at higher Re_s, the shape of bubbles and drops becomes elongated. Under these conditions, the drag coefficient of elongated viscous spheres is a function of the variables that appear in the above correlations as well as of the amount of surface deformation, which is expressed by the eccentricity. In this case, correction functions must be used with the above correlations for the eccentricity, such as the one derived by Harper (1972). Experimental evidence by Winnikow and Chao (1966) on drops moving in liquids shows that the drops will remain spherical when the Bond number, Bo, is less than or equal to 0.2. Accordingly, water droplets in air will maintain their spherical shape at values of Re_s up to 470. For drops of substances with high surface tension (liquid metals), the corresponding Re_s would be significantly higher: Mercury drops in air maintain their spherical shape up to $Re_s = 1,150$.

1.3.3 Drag on Irregular Particles

Experiments have shown that the drag coefficients for non-spherical particles are different than those of spheres. The irregular shape of these particles is the main reason that modifies the values of the drag coefficient. Shape factors have been proposed as parameters for the quantification of the effects of the irregularity of particles, bubbles, and drops. In addition, irregular particles have more than one lengthscales, which are often significantly different. The volume-equivalent and the area-equivalent diameters of equivalent spheres, d_n and d_A, respectively, are often used as the characteristic lengths of irregularly shaped particles in correlations for the drag coefficient:

$$d_n = \sqrt[3]{6V/\pi}, \text{ and } d_A = \sqrt{4A_p/\pi}, \tag{1.35}$$

where V is the total volume and A_p is the projected area of the particle in the direction of the flow. Since it is impossible to derive analytical or numerical results for the multitude of the shapes of irregular particles, this subject has been dominated by experimental studies and empirical correlations. Among the earlier studies, Wadell (1933) defined a shape factor, Ψ, for the drag

$$\Psi = \frac{A_s}{A} = \frac{d_n^2}{d_A^2} = \frac{\pi^{1/3}(6V)^{2/3}}{A_p} \tag{1.36}$$

and suggested a simple correlation for the drag coefficient, C_D. Pettyjohn and Christiansen (1948), Haider and Levenspiel (1989), and Hartman and Yates (1993) suggested that when the particles are very elongated and the correction factors are large, the circularity or sphericity, c, be used:

$$c = \frac{\pi d_A}{P_P}, \tag{1.37}$$

where P_p is the projected perimeter of the particle in the direction of motion. Other studies on this subject are by Lasso and Weidman (1986), Haider and Levenspiel (1989), Chhabra et al. (1995), and Madhav and Chhabra (1995) who presented useful correlations for the drag coefficients of several classes of irregular particles. Tran-Cong et al. (2004) summarized these correlations and, based on their own experimental results, offered a new and more accurate correlation for irregularly shaped particles, which is as follows:

$$C_D = \frac{24}{Re_s}\frac{d_A}{d_n}\left[1 + \frac{0.15}{\sqrt{c}}\left(\frac{d_A}{d_n}Re_s\right)^{0.687}\right] + \frac{0.42}{\sqrt{c}\left[1 + 4.25 \times 10^4\left(\frac{d_A}{d_n}Re_s\right)^{-1.16}\right]}. \tag{1.38}$$

The last equation is valid in the ranges $0.15 < Re_s < 1{,}500$, $0.80 < d_A/d_n < 1.50$ and $0.4 < c < 1.0$, which cover most of the engineering applications.

1.3.4 Blowing Effects

Many of the applications for particulate heat transfer pertain to the evaporation or combustion of droplets and particles. The mass transfer from the surface of a sublimating particle or an evaporating droplet causes significant changes in the gaseous flow field around it. The effects of these changes include a reduction of the drag coefficient from the steady values given by the standard drag curves of Figs. 1.2 and 1.3. Also, the mass flux from the particle or droplet to the carrier fluid causes two significant effects close to the interface:

1. The change of the carrier fluid viscosity, which occurs because of temperature differences and/or because of the carrier fluid composition changes.

2. The regression of the surface of the droplet and the development of a radial velocity field, which is associated with the flow of the vapor from the surface to the carrier fluid. This is often called the *Stefan convection*.

Yuen and Chen (1976) conducted experiments on the drag force of evaporating drops and on the effect of the change of the viscosity of the carrier fluid. They concluded that the reference viscosity of the carrier fluid, which is best to be used in calculations with evaporating drops, is the *film viscosity*, defined as:

$$\mu_{\mathrm{m}} = \mu_{\mathrm{s}} + \frac{1}{3}(\mu_{\infty} - \mu_{\mathrm{s}}), \tag{1.39}$$

where μ_{∞} is the viscosity of the fluid far from the sphere and μ_{s} the gaseous viscosity on the surface of the sphere. This equation is sometimes referred to as the "1/3 rule" for the film properties and is used with the other transport coefficients, thermal conductivity and diffusivity. The experimental data and analysis on the density and viscosity averages by Lerner et al. (1980) confirmed this relationship. The last study also suggested that the standard drag curve with the above correction for the viscosity could be applied to slightly ellipsoidal drops if the volume-equivalent diameter of the ellipsoid, d_{n}, which is defined in Eq. (1.35), is used and the two axes of the ellipsoid are within 10% of each other.

All the blowing effects are associated with the mass or heat transfer at the interface of particles or drops. In order to model these effects, two "blowing coefficients" have been defined for the mass and heat transfer processes as follows:

$$B_{\mathrm{M}} = \frac{Y_{\mathrm{s}} - Y_{\infty}}{1 - Y_{\mathrm{s}}}. \tag{1.40}$$

and

$$B_{\mathrm{H}} = \frac{h_{\infty} - h_{\mathrm{s}}}{h_{\mathrm{fg}}^{\mathrm{eff}}}. \tag{1.41}$$

The symbols Y and h denote the mass fraction of a chemical species in the carrier fluid and the enthalpy of the fluid. The subscripts s, ∞, and fg denote the surface of the sphere, a distance far from the sphere, and latent heat, respectively. The effective latent heat of vaporization in the denominator of the last expression is the sum of the latent heat of the vapor at the drop surface and the sensible heat that is conducted to the interior of the drop. In most cases the sensible heat is very low in comparison to the latent heat and $h_{\mathrm{fg}}^{\mathrm{eff}} \approx h_{\mathrm{fg}}$.

Chiang et al. (1992) conducted a numerical study and derived a correlation for the drag coefficients of evaporating drops, valid in the range $30 < Re_{\mathrm{s}} < 200$:

$$C_{\mathrm{D}} = \frac{24.432}{(1 + B_{\mathrm{H}})^{0.27}(Re_{\mathrm{m}}/2)^{0.721}}. \tag{1.42}$$

The Reynolds number Re_m in the last expression is defined in terms of the gas-film viscosity of Eq. (1.39) and the free-stream gas density, $\rho_{f\infty}$.

When the radial velocity of blowing, V_n, is known or may be calculated, Clift and Lever (1985) derived from numerical results the following relationship for the drag coefficient:

$$C_D = \frac{24}{Re_s} \frac{1 + 0.545Re_s + 0.1Re_s^{0.5}(1 - 0.03Re_s)}{1 + A\,Re_n^B}, \qquad (1.43)$$

where Re_n is the "blowing Reynolds number," $Re_n = 2\alpha V_n/\nu$, and the functions A and B are correlated to the particle Reynolds number as follows:

$$A = 0.09 + 0.077\exp(-0.4Re_s) \quad \text{and} \quad B = 0.4 + 0.77\exp(-0.04Re_s). \quad (1.44)$$

These correlations and the pertinent studies suggest that the drag coefficients of burning and evaporating droplets differ little from the corresponding values derived in the absence of mass transfer. The correlations may be perceived as corrections to the standard drag curve for spheres that are caused by the different composition of the gaseous boundary layer and the vapor convection from the sphere.

1.3.5   Other Effects on the Steady Drag Coefficients

While the viscosities of the carrier fluid and the sphere are the most important parameters, other effects also affect significantly the steady drag coefficients of particles, bubbles, and drops. Among these parameters are the following:

- Carrier fluid turbulence
- Compressibility or rarefaction effects
- Effect of surfactants
- Proximity to solid or permeable boundaries

A more thorough exposition of these effects and the suggested corrections for the drag coefficients may be found in the monograph by Michaelides (2006) or the review paper by Michaelides (2003).

1.3.6   Transient Flow

Many applications of dispersed multiphase flow involve time-dependent effects. When the characteristic time of the fluid transients, τ_f, is significantly higher than the timescale of the particle, τ_M, ($\tau_f \gg \tau_M$), the process may be considered as quasi-static and the steady equation of motion may be used. However, when the characteristic time of the transients in the carrier flow is of the same order of magnitude as the characteristic time of the dispersed phase, a transient equation must be used for the exchange of momentum between the carrier fluid and the dispersed phase. In the other extreme, when $\tau_M \gg \tau_f$, the dispersed phase does not respond to the fluctuations of the carrier fluid. In the case $\tau_M \sim \tau_f$, exact analytical expressions for transient flows have been derived for creeping flow conditions ($Re_s \ll 1$) and asymptotic expressions have been derived for finite but small Reynolds numbers ($Re_s < 1$) only for solid spheres. Semi-empirical expressions, which emanate from a combination of experimental data and analysis, have also been developed and frequently used at high Re_s.

1.3.6.1 Creeping Flow ($Re_s \ll 1$)

Boussinesq (1885) and Basset (1888) derived independently the first equation for the transient hydrodynamic force on a solid sphere at $Re_s \ll 1$. Maxey and Riley (1983) performed a mathematically rigorous analysis for a rigid sphere in an arbitrary nonuniform flow field, whose velocity vector is given by the functional relationship $u_i(x_i,t)$. Their final form of the equation of motion is:

$$
m_s \frac{dv_i}{dt} = -\frac{1}{2} m_f \frac{d}{dt}\left(v_i - u_i - \frac{\alpha^2}{10} u_{i,jj}\right) - 6\pi\alpha\,\mu_f \left(v_i - u_i - \frac{\alpha^2}{6} u_{i,jj}\right)
$$
$$
- \frac{6\pi\alpha^2\,\mu_f}{\sqrt{\pi\,v_f}} \int_0^t \frac{\frac{d}{d\tau}\left(v_i - u_i - \frac{\alpha^2}{6} u_{i,jj}\right)}{\sqrt{t - \tau}} d\tau + (m_s - m_f)\,g_i + m_f \frac{Du_i}{Dt}, \quad (1.45)
$$

where m_s is the mass of the sphere and m_f is the mass of the fluid that occupies the same volume as that of the sphere; the repeated index (jj) denotes the Laplacian operator, and the derivative D/Dt is the total Lagrangian derivative following the sphere. The Laplacian terms $u_{i,jj}$ arise from the nonuniformity of the velocity field of the carrier fluid and are sometimes called the "Faxen terms" (Faxen 1922). All the spatial derivatives are evaluated at the center of the sphere. The left-hand side of the last equation represents the acceleration of the sphere. Of the terms in the right-hand side, the first represents the added mass, which is the mass of the fluid that must be accelerated with the sphere; the second is the steady drag on the sphere; and the third is the history term, which is sometimes called the "Bassett term." The last two terms in the right-hand side are the gravitational, or body, force and the Lagrangian acceleration term, caused by the acceleration of the fluid. The Faxen terms scale as α^2/L^2, where L is the macroscopic characteristic length of the fluid velocity. In most practical applications of dispersed multiphase flows, $\alpha/L \ll 1$, and the Faxen terms are small enough to be neglected.

It must be noted that the derivation of Eq. (1.45) is based on the following assumptions:

- Spherical shape
- Infinite fluid domain initially undisturbed

- No rotation
- Rigid sphere ($\mu_f/\mu_s \ll 1$)
- Zero initial relative velocity
- Negligible inertia effects ($Re_s \ll 1$)

If the initial relative velocity is different than zero, the following expression/ correction replaces the history term of equation (1.45):

$$\frac{6\pi a^2 \mu_f}{\sqrt{\pi \nu_f}} \int_0^t \frac{\frac{d}{d\tau}\left[v_i - u_i - \frac{a^2}{6} u_{i,jj}\right]}{\sqrt{t - \tau}} d\tau + \frac{6\pi a^2 \mu_f [v_i(0) - u_i(0)]}{\sqrt{\pi \nu_f t}}, \tag{1.46}$$

where the quantity $v_i(0) - u_i(0)$ is the initial relative velocity (at time $t = 0$). A discussion on this initial condition and on the history terms in general may be found in Michaelides (2003) and (2006). A conversion of the implicit in v_i integrodifferential equation (1.45) to a second-order explicit expression, which saves a significant amount of computational time, may also be found in these sources as well as in Vojir and Michaelides (1994).

In the case of viscous spheres (drops and bubbles), there are two timescales for the motion of the fluid inside the sphere and the motion of the external fluid. An analytical expression for the equation of motion, similar to Eq. (1.45), is impossible to be derived in the time domain. Galindo and Gerbeth (1993) were the first to derive a correct expression for the hydrodynamic force on a viscous sphere under creeping flow conditions. Similarly, a general expression for the hydrodynamic force with slip at the interface may only be obtained in the Laplace domain. A general case and special cases with slip have been derived by Michaelides and Feng (1995) and later by Feng et al. (2012). The latter study covers several cases at the interface of continuum and molecular dynamics and is applicable to nanoparticles. It includes the Knudsen number as a parameter and applies to solid as well as viscous spheres.

Parmar et al. (2011) examined the effects of fluid compressibility and performed an analytical study on the motion of a solid sphere in compressible flow at $Re_s \ll 1$ with applications to shock wave and particle interactions. They expressed the effects of the fluid compressibility by a correction function, which depends on the Knudsen number and the ratio of the bulk to the dynamic viscosities. In a companion study, Ling et al. (2011a, b) determined the effects of the unsteady terms in the dispersion of small particles in shock waves. Their results show that the transient term contributions to the hydrodynamic force and heat transfer coefficient of the particles are significant. At the early stages of the blast wave, when particles are traveling in the expansion fan, the gas density surrounding the particles is larger. This makes the transient components of the hydrodynamic force and the rate of heat transfer to be of the same order of magnitude as the steady terms. Consequently, the error from neglecting the transient terms in such applications may become significant.

1.3.6.2 Finite but Small Reynolds Numbers

When advection and inertia effects become important, only asymptotic solutions may be derived for the transient equation of motion, even for solid, spherical particles. Sano (1981) used an asymptotic analysis and derived an expression for the transient hydrodynamic force acting on a rigid sphere at small but finite values of the Reynolds number, when the sphere undergoes a step change of velocity. Mei et al. (1991) conducted a numerical study on the motion of a rigid spherical particle in the range $0 < Re_s < 50$ and showed that the fluid advection causes a faster decay of the history term than the conventional $t^{-1/2}$ rate of Eq. (1.45). Mei and Adrian (1992) obtained an analytical solution for the motion of a solid sphere, which is valid at very low frequencies ($Sl \ll Re_s < 1$). Lovalenti and Brady (1993a, b) followed Sano's analytical method and derived a more general expression for the hydrodynamic force on a sphere undergoing arbitrary motion. One of the principal conclusions of this study is that the unsteady terms of the equation of motion decay faster when flow advection is significant, because the fluid vorticity around the particle is advected faster to an outer region.

1.3.6.3 High Reynolds Numbers: Semi-empirical Expressions

While all the known analytical expressions for the hydrodynamic force on particles, bubbles, and drops apply to low Reynolds numbers, several engineering applications of the transport processes occur at higher ranges of this parameter: Slurry transport and practical pneumatic conveying systems operate in the range $10^1 < Re_s < 10^3$; drops in combustion processes may reach Reynolds numbers up to 10^3; bubble columns in chemical processes operate in a range from 0 to 10^3; and particulate flows in the environment may reach values of Re_s up to 10^4. Since there is no applicable theory for $Re_s > 1$, experimental data and empirical correlations have been used for the calculation of the transient hydrodynamic force. Among these, the most frequently used is the one by Odar and Hamilton (1964): They essentially treated the three terms in Eq. (1.45) as separate, independent forces and associated empirical correction factors, C_1, Δ_A, and Δ_H, with each one of them. The final expression of the unsteady equation of motion, which does not include the Faxen terms, $u_{i,jj}$, is as follows:

$$m_s \frac{dv_i}{dt} = -\frac{\Delta_A}{2} m_f \frac{d}{dt}(v_i - u_i) - 6C_1\pi\alpha\mu_f(v_i - u_i) - \frac{6\Delta_H\pi\alpha^2\mu_f}{\sqrt{\pi\nu_f}}$$

$$\times \int_0^t \frac{\frac{d}{d\tau}(v_i - u_i)}{\sqrt{t - \tau}} d\tau + (m_s - m_f)g_i + m_f\frac{Du_i}{Dt}, \tag{1.47}$$

where C_1 is the term in parenthesis in Eq. (1.28) and the parameters Δ_A and Δ_H are defined in terms of the acceleration factor Ac:

$$\Delta_A = 1.05 - \frac{0.066}{0.12 + Ac^2}, \Delta_H = 2.88 + \frac{3.12}{(1+Ac)^3}, \text{ with } Ac = \frac{|u_i - v_i|^2}{2a\left|\frac{dv_i}{dt}\right|}, \quad (1.48)$$

Al-taweel and Carley (1971) performed independently experiments for Δ_A and Δ_H and derived different correlations for the two coefficients. Karanfilian and Kotas (1978) suggested constant values for Δ_A and Δ_H of 0.5 and 6, respectively. A thorough discussion of the subject and the semi-empirical transient equation of motion are given by Michaelides (2003, 2006), and a reinterpretation of the original results by Odar and Hamilton is given by Michaelides and Roig (2011).

1.3.7 Transverse Forces and Lift Effects

The drag force always acts in the direction of motion of the particles. Particle rotation and fluid velocity gradients (shear) combined with finite relative velocity between the fluid and particle will induce a transverse component in the hydrodynamic force on the particle, which is often called *the lift force*. When a rigid, spherical particle traverses a fluid with a relative velocity, and also rotates with respect to the flow, a transverse pressure difference is developed on the surface of the sphere, which results in the so-called Magnus force (Magnus 1861). The Magnus force is given by the following expression:

$$\vec{F}_{LM} = \pi\alpha^3 \rho_f \vec{\Omega} \times (\vec{v} - \vec{u}), \quad (1.49)$$

where Ω is the relative rotation of the particle with respect to the fluid and the symbol $\times$ represents the vector product (cross product) of the two vectors. The direction of the Magnus force is perpendicular to the plane of the relative velocity and the axis of rotation.

The lift force is the consequence of the sideways pressure difference induced because of the streamline asymmetry, which stems from the rotation of the sphere and the no-slip boundary condition at the interface. The Magnus force is not a consequence of the fluid viscosity, and therefore, it affects the motion of particles in both viscous and inviscid fluids. As it happens with the steady drag force, the magnitude of the lift force is expressed as a function of a dimensionless *lift coefficient*, C_{LM}, and the latter is correlated with experimental data. The following expression, derived by Oesterle and Bui Dinh (1998), is recommended for the lift coefficient of a sphere rotating in an infinite fluid:

$$C_{LM} = 0.45 + \left(\frac{Re_{rot}}{Re_s} - 0.45\right) \exp(-0.05684 Re_{rot}^{0.4} Re_s^{0.3}) \quad \text{for} \quad Re_{rot} < 140,$$

$$(1.50)$$

where Re_{rot} is a Reynolds number based on the relative rotational speed of the sphere and was defined in Eq. (1.3). The lift coefficient is in general a monotonically decreasing function of Re_s and, in general, increases with the dimensionless *rotation parameter*, $2\alpha|\vec{\Omega}|/|\vec{u} - \vec{v}|$. The latter is a dimensionless measure of the rotational speed of the particle. Experiments at low values of the rotation parameter suggest that C_{LM} may become negative (Tanaka et al. 1990). This is most likely due to a higher relative velocity on one side of the sphere and the premature transition of the boundary layer to turbulence.

When a particle is present in a region of fluid shear, there is a de facto relative rotation between the fluid and the particle, which induces a transverse force. Saffman (1965, 1968) considered the case of a very small sphere in a shear flow at the limit of vanishing Reynolds number (creeping flow). He derived an expression for this force whose magnitude is:

$$\vec{F}_{LS} = \frac{6.46\alpha^2\sqrt{\rho_f\mu_f}}{\sqrt{|\vec{\gamma}|}}(\vec{u} - \vec{v}) \times \vec{\gamma}, \tag{1.51}$$

where γ is the fluid velocity shear evaluated at the center of the sphere. The direction of the lift force is in the perpendicular direction to the plane defined by the relative velocity vector and the shear vector. It must be noted that the lift force on a sphere is, in general, very much weaker than the longitudinal drag force. However, the transverse lift force plays a dominant role in the lateral migration of bubbles, drops, and particles toward the walls of cylindrical conduits as well as in dispersion, because it is the principal driving force in the lateral direction. This weak transverse force contributes significantly to the radial diffusion and dispersion, wall deposition, mixing, and separation processes.

Among the recent developments on the lift force exerted by a viscous fluid on a sphere, Tsuji et al. (1985) measured experimentally the shear-induced lift on bubbles and concluded that Eq. (1.51) yields satisfactory results on the magnitude of the force. McLaughlin (1991) extended the theoretical analysis by Saffman (1965) to higher values of Re_s and derived a correction to this expression, which shows that the magnitude of F_{LS} decreases with increasing Re_s.

Dandy and Dwyer (1990) conducted a numerical study and derived results for the lift force exerted on a sphere by the flow shear and finite Re_s. These results were reduced to a useful correlation by Mei (1992) who corrected the original Saffman expression accordingly. Mei's (1992) proposed correction is as follows:

$$\vec{F}_{LS} = \left[\frac{6.46\alpha^2\sqrt{\rho_f\mu_f}}{|\vec{\gamma}|}(\vec{u} - \vec{v}) \times \vec{\gamma}\right]$$
$$\times \left[\left(1 - 0.3314\sqrt{\frac{Re_\gamma}{2Re_s}}\right)\exp\left(-\frac{Re_s}{10}\right) + 0.3314\sqrt{\frac{Re_\gamma}{2Re_s}}\right], \tag{1.52}$$

for $Re_s < 40$ and

$$\vec{F}_{LS} = \left[\frac{6.46\alpha^2 \sqrt{\rho_f \mu_f}}{|\vec{\gamma}|} (\vec{u} - \vec{v}) \times \vec{\gamma} \right] \left[0.0524 \sqrt{\frac{Re_\gamma}{2}} \right], \qquad (1.53)$$

for $Re_s > 40$. These expressions are recommended to be used in the range $0.01 < Re_\gamma/Re_s < 0.8$. Mei (1992) also showed that this empirical equation fits well McLaughlin's results.

It must be noted that, because the lift force is weak and difficult to measure, there is no widespread agreement on its magnitude, especially for viscous spheres, such as bubbles and drops. The experimental results by Sridhar and Katz (1995) suggest that, at least for bubbles, the magnitude of the lift force is higher than the values predicted by the above two expressions. Another experimental study on bubbles by Tomiyama et al. (1999) concluded that the lift coefficient of larger bubbles depends on the surface tension and, hence, on the Eötvös number. Tomiyama et al. (1999) also concluded that rising bubbles tend to accumulate close to the walls if their radii are less than 6 mm, while bigger bubbles with $\alpha > 6$ mm tend to migrate toward the center of the conduit.

The transverse force on particles is also influenced by the presence of other particles, which distort the flow field. Feng and Michaelides (2002) studied the shear-induced lift on a sphere attached to a boundary in the presence of other spheres in the flow field. Their results show that the instantaneous hydrodynamic force exerted by the suspension flow on the attached solid sphere is increased by a factor of 2–4 when the suspended spheres pass in close proximity. As a result of this type of interaction, a particle that lies on a horizontal plane may be lifted and join the suspension flow without physical collisions with other particles. Such particle–particle interactions in shear flows would also decrease the rate of sedimentation as well as the rate of surface erosion.

During evaporation and sublimation, there is an influence of the mass transfer from the sphere on the lift force. Kurose et al. (2003) determined that in linear shear flow, the outflow velocity from a sphere acts in a way to push the sphere toward the lower velocity side. Thus, a negative lift is developed at higher Reynolds numbers. This tends to counteract the positive lift on the sphere, which is directed toward the higher-velocity side. According to the numerical study by Kurose et al. (2003), the diffusion and reaction rates are strongly affected by the outflow velocity because of the deformation of the vortices, which develop behind the evaporating sphere. This study implies that surface evaporation and/or chemical reactions are of high importance to the lateral motion of particles and drops undergoing combustion processes.

1.4 Heat and Mass Transfer

Most of the analytical techniques on the subject of heat transfer are based on the seminal work by Jean-Baptiste Joseph Fourier (Fourier 1822). His treatise, which was preceded by seven shorter articles, has been supplemented by numerous experimental studies that have provided semi-empirical correlations for the rate of heat transfer. For this reason, the subject of heat transfer from spheres is based primarily on experimental correlations or, in the recent past, on results from numerical studies.

At first, it must be noted that the continua governing equations for the heat transfer and for the mass transfer are similar. The solution of one equation is also the solution for the other equation by simple substitution of the corresponding variables and dimensionless numbers. For brevity, we will refer to equations and results for the heat transfer, but it must be noted that all the heat transfer results, analytical, experimental, and computational, are also applicable to the process of mass transfer. The general transient energy equation in differential form is

$$k_f \nabla^2 T = \rho_f c_f \left(\frac{\partial T}{\partial t} + \vec{u} \cdot \vec{\nabla} T \right). \tag{1.54}$$

or in dimensionless form:

$$\nabla^2 T^* = \frac{\partial T^*}{\partial t^*} + \mathrm{Pe}_f \cdot \nabla T^*, \tag{1.55}$$

where the symbol * denotes dimensionless variables. Time is made dimensionless in Eq. (1.55) using the normalizing timescale $\rho_f c_f L^2 / k_f$.

It is apparent that an important parameter for the study of convective heat and mass transfer from spheres is the Peclet number, which accounts for the heat advected by the fluid. The Peclet number is analogous to the Reynolds number in the equation of motion and is defined in Eq. (1.4). It may be seen in this equation that several Peclet numbers may be defined (a) for a particle, Pe_s; (b) for the fluid, Pe_f; (c) for shear or rotational effects, Pe_y; and (d) the Peclet number that is pertinent to the mass transfer, Pe_m. The Nusselt number and the Sherwood number are also important dimensionless numbers, and they are used for convective heat and mass transfer, respectively.

1.4.1 Steady, Stokesian Heat Transfer for Spheres

The case of creeping flow or Stokes flow, which was presented in Sect. 1.3.1, implies $Re_s \ll 1$. Because of the relationship $Pe_s = Re_s * Pr$, and unless Pr for the sphere is very large, as is the case of some organic oils, this condition also implies that $Pe_s \ll 1$, or at least that $Pe_s < 1$.

Heat convection is comprised of two parts: conduction and advection. When $Pe_s \ll 1$, conduction dominates and advection is insignificant. In this case, one typically neglects the effects of advection and treats the conduction part of the equation alone. Thus, the solution of the governing equation for steady conduction from a sphere may be obtained. This solution yields the following expression for the Nusselt number:

$$Nu = 2. \tag{1.56}$$

When Pe_s is small but finite, $Pe_s < 1$, the solution of the governing energy equation may be done asymptotically. Acrivos and Taylor (1962) conducted a study on the heat transfer from a sphere, which is analogous to the study by Proudman and Pierson (1956) for the equation of motion. They implicitly assumed Stokesian flow around a sphere and derived an asymptotic heat transfer solution, which is valid for $Pe_s < 1$. With later corrections for the coefficients (Acrivos 1980; Leal 1992), the expression for the Nusselt number at steady conditions is as follows:

$$Nu = 2 + \frac{Pe_s}{2} + \frac{1}{4}Pe_s{}^2 \ln\frac{Pe_s}{2} + 0.2073Pe_s{}^2 + \frac{1}{16}Pe_s{}^3 \ln\frac{Pe_s}{2}. \tag{1.57}$$

The last equation is applicable in the ranges $Re_s \ll 1$ and $Pe_s < 1$.

Acrivos and Taylor (1962) also proved that the functional relationship $Nu(Pe_s)$, as obtained assuming Stokesian flow, is less sensitive to an increase of Re_s than the corresponding functional relationship for the drag coefficient, $C_D(Re_s)$. Therefore, it is generally accepted that Eq. (1.57) is valid not only under the creeping flow conditions, $Re_s \ll 1$, but also when Re_s is finite but small ($Re_s < 1$). This was confirmed by other studies, including that of Brunn (1982).

Acrivos and Goddard (1965) derived an asymptotic solution for a solid sphere at high Pe_s assuming a Stokesian velocity distribution around the sphere. Their expression, which is valid for $Pe_s > 5$, may be written as follows:

$$Nu = 1.249 \left(\frac{Pe_s}{2}\right)^{1/3} + 0.922. \tag{1.58}$$

In the case of viscous spheres, drops, or bubbles, the viscosity ratio, λ, and the internal motion affect the heat transfer process. Levich (1962) derived an asymptotic, first-order solution for a liquid sphere at very large Pe_s ($Pe_s \gg 1$) under the creeping flow conditions ($Re_s \ll 1$):

$$Nu = \sqrt{\frac{4Pe_s}{3\pi(1 + \lambda)}}. \tag{1.59}$$

More recently, Feng and Michaelides (2000b) solved numerically the energy equation of a viscous sphere at creeping flow conditions. They correlated their numerical data with the following equation:

$$Nu = 1.49 Pe_s^{0.322 + \frac{0.113}{0.361\lambda + 1}}. \tag{1.60}$$

It must be noted that the implicit conditions for the use of the last three expressions are $Re_s \ll 1$ (or at least $Re_s < 1$) and $Pe_s \gg 1$. These conditions are satisfied only for spheres with high Prandl numbers. Drops of several organic liquids including gasoline and engine oil satisfy these conditions.

1.4.2   Inertia Effects, Higher Re_s

The empirical correlations of the experimental data by Ranz and Marshall (1952) and Whitaker (1972) have been proven to accurately describe the convective heat transfer from small solid spheres:

$$Nu = 2 + 0.6 \, Re_s^{0.5} Pr^{0.33}, \tag{1.61}$$

and

$$Nu = 2 + \left(0.4 \, Re_s^{1/2} + 0.06 \, Re_s^{2/3} \right) Pr^{0.4}. \tag{1.62}$$

These correlations may be used up to $Re_s = 10^4$ and are valid for solid spheres only.

Regarding viscous spheres, Feng and Michaelides (2000a, 2001b) conducted numerical studies on the subject and derived useful correlations, with Re_s and Pe_s as independent variables. The first study pertains to high Re_s and any Pe_s (Feng and Michaelides 2000a) and the second to any values of Re_s and Pe_s (Feng and Michaelides 2001b). Their results in correlation form may be summarized as follows:

A. In the range $0 < Re_s < 1$, the general expression for the Nusselt number is

$$Nu(\lambda, Pe_s, Re_s) = \left[\frac{0.651}{1 + 0.95\lambda} Pe_s^{1/2} + \frac{0.991\lambda}{1 + \lambda} Pe_s^{1/3} \right] [1 + f(Re_s)]$$
$$+ \left[\frac{1.65(1 - f(Re_s))}{1 + 0.95\lambda} + \frac{\lambda}{1 + \lambda} \right], \tag{1.63}$$

where the function $f(Re_s)$ is defined as follows:

$$f(Re_s) = \frac{0.61 Re_s}{Re_s + 21} + 0.032. \tag{1.64}$$

B. As in the case of the expression of the drag coefficient in Sect. 1.3.2, for higher Re_s, the analysis of the heat transfer data for viscous spheres revealed that the best correlations are obtained when the general expression for the Nusselt number is given in terms of the following three functions, which pertain to values of the viscosity ratio, λ, equal to 0, 2, and infinity (very large):

B1. The correlation for inviscid spheres, typically bubbles, $(\lambda = 0)$:

$$Nu(0, Pe_s, Re_s) = 0.651 \, Pe_s^{1/2} \left(1.032 + \frac{0.61 Re_s}{Re_s + 21} \right)$$

$$+ \left(1.60 - \frac{0.61 Re_s}{Re_s + 21} \right). \tag{1.65}$$

B2. The correlation for solid spheres $(\lambda = \infty)$:

$$Nu(\infty, Pe_s, Re_s) = 0.852 \, Pe_s^{1/3} \left(1 + 0.233 Re_s^{0.287} \right) + 1.3$$

$$- 0.182 Re_s^{0.355}. \tag{1.66}$$

B3. The corresponding function for a sphere with viscosity ratio $\lambda = 2$, which was derived as follows from the numerical results:

$$Nu(2, Pe_s, Re_s) = 0.64 Pe_s^{0.43} \left(1 + 0.233 Re_s^{0.287} \right) + 1.41 - 0.15 Re_s^{0.287}. \tag{1.67}$$

Accordingly, the final correlations for the heat transfer coefficients are given by the following expressions:

A. In the range $0 \leq \lambda < 2$:

$$Nu(Pe_s, Re_s, \lambda) = \frac{2 - \lambda}{2} Nu(Pe_s, Re_s, 0) + \frac{4\lambda}{6 + \lambda} Nu(Pe_s, Re_s, 2). \tag{1.68}$$

B. In the range $2 < \lambda \leq \infty$:

$$Nu(Pe_s, Re_s, \lambda) = \frac{4}{\lambda + 2} Nu(Pe_s, Re_s, 2) + \frac{\lambda - 2}{\lambda + 2} Nu(Pe_s, Re_s, \infty). \tag{1.69}$$

The last two correlations are valid for $Pe_s > 10$. For smaller values of Pe_s, it was not possible to obtain a simple correlation of the numerical results with any satisfactory degree of accuracy. For this reason, in the range $0 < Pe_s < 10$, it is recommended to use the numerical results in the original publication (Feng and Michaelides 2001b), which are given in tabular form.

As in the case of the hydrodynamic force on a viscous sphere, it was determined that, for fixed values of Re_s and viscosity ratio, λ, the variations of the density ratio,

ρ_s/ρ_f, have only a minimal effect on the external flow field. When one considers the governing equation for the heat and mass transfer processes and the pertinent boundary conditions, one will conclude that the density ratio (or equivalently the internal Reynolds number, Re_i) does not affect significantly the corresponding transport coefficients, h_c or h_M, which pertain to the heat and mass convection in the outer fluid and, consequently, the Nusselt or Sherwood numbers. This was verified by Feng and Michaelides (2001a, b) in extensive numerical computations for both the hydrodynamic force and for the rate of heat transfer. The results of these computations show conclusively that the influence of the density ratio on the heat transfer coefficient is less than 0.1%. This value is of the same order of magnitude as the numerical uncertainty of the computations and much lower than the required accuracy for any engineering calculations.

1.4.3   Blowing Effects

Blowing effects are important for burning droplets when the timescale of mass transfer from the drop is of equal or lesser order of magnitude than the timescale of energy transfer. As in the case of momentum transfer, corrections to the heat transfer coefficient have been developed. These corrections take into account the change of the properties of the gaseous boundary layer and the phase change on the surface of the sphere. The two *blowing factors*, B_H and B_M, which were defined in Sect. 1.3.4 account for the heat and mass transfer effects on the surface of the sphere. Since the origin of the two blowing factors is the radial mass transfer from the surface of the sphere to the carrier fluid, it is evident that the two are not independent. Abramzon and Sirignano (1989) conducted an analytical study on the evaporation of drops and derived expressions for the two blowing factors that are useful in engineering calculations. For a burning drop in air, they derived the following relationship:

$$B_H = (1 + B_M)^n - 1, \tag{1.70}$$

where the exponent n is the ratio:

$$n = \frac{c_{pF}}{c_{pf}} \frac{1}{Le} \frac{1 + 0.424 \frac{\sqrt{Re_s/2}}{F(B_M)}}{1 + 0.424 \frac{\sqrt{Re_s/2}}{F(B_H)}}. \tag{1.71}$$

In the last equation, c_{pF} is the specific heat of the vapor that emanates from the drop, c_{pf} is the specific heat of the carrier gas, Le is the Lewis number, and F is a function of the corresponding blowing factor:

$$F(B) = (1 + B)^{0.7} \frac{\log(1 + B)}{B}. \tag{1.72}$$

Abramzon and Sirignano (1989) obtained semi-analytical expressions for the Nusselt and Sherwood numbers of an evaporating or burning drop. Later, Chiang et al. (1992) improved on that study by relaxing some of the assumptions and conducting numerical computations. Chiang et al. (1992) derived the following correlations for the heat and mass transfer coefficients:

$$Nu = 1.275(1 + B_H)^{-0.678} Re_m{}^{0.438} Pr_m{}^{0.619},\tag{1.73}$$

and

$$Sh = 1.224(1 + B_M)^{-0.568} Re_m{}^{0.385} Sc_m{}^{0.492}.\tag{1.74}$$

The Reynolds and Schmidt numbers, Re_m and Sc_m, are calculated using the film properties of the carrier gas and the vapor using the "1/3 rule" of Eq. (1.39). The free-stream gas density $\rho_{f\infty}$ is used to define Re_m. One may combine the above correlations and obtain the following expressions for the rate of heat and mass transfer from the surface of a spherical drop:

$$
\begin{aligned}
\dot{Q} &= 2.55\,\pi\alpha k_f(T_s - T_\infty)(1 + B_H)^{-0.678} Re_m{}^{0.438} Pr_m{}^{0.619} \\
\dot{m} &= 2.448\,\pi\rho_f D_f(Y_s - Y_\infty) B_M (1 + B_M)^{-0.568} Re_m{}^{0.385} Sc_m{}^{0.492}.
\end{aligned}\tag{1.75}
$$

1.4.4 Transient Effects

1.4.4.1 At Creeping Flow Conditions

Michaelides and Feng (1994) conducted a study on the transient energy equation for spheres, which is analogous to the transient equation of motion by Maxey and Riley (1983). They calculated analytically the contributions of the far-away temperature field and the near-field to the sphere and obtained the total time-dependent heat transfer rate. For a rigid, isothermal sphere, in a time-variable, and nonuniform fluid temperature field $T_f(x_i,t)$, they derived the following general energy equation for a solid sphere:

$$
m_s\,c_s\,\frac{dT_s}{dt} = -\,m_f\,c_f\,\frac{DT_f}{Dt} - 4\pi\alpha\,k_f\left[T_s - T_f - \frac{1}{6}\alpha^2 T_{f,ii}\right] - 4\pi\,\alpha^2\,k_f
$$

$$
\times \int_0^t \frac{\frac{d}{d\tau}\left[T_s - T_f - \frac{1}{6}\alpha^2 T_{f,ii}\right]}{[\pi a_f(t - \tau)]^{1/2}}\,d\tau,\tag{1.76}
$$

where c_f and c_s are the specific heat capacities of the fluid and the sphere, respectively;[3] α_f in the denominator of the last term is the thermal diffusivity of the fluid, which is equal to $k_f/\rho_f c_{pf}$; τ is a dummy variable with units of time; and, the repeated index jj denotes the Laplacian operator. As with the corresponding terms of the equation of motion, the terms $T_{f,jj}$ take into account the spatial nonuniformity of the fluid temperature field and scale as α^2/L^2.

The first term in the r.h.s. of Eq. (1.76) represents the contribution of the outer temperature field to the rate of heat transfer. The second term is the usual Fourier conduction term with the addition of the Laplacian term. The last term, which was named the *history term*, is analogous to the history term of the equation of motion of a sphere. The physical origin of the history term is the diffusion of the temperature gradients in the fluid around the solid sphere and represents the additional fluid-particle energy exchange that accompanies the gradient diffusion process.

The existence of the history term has been alluded to by Carslaw and Jaeger (1947) who obtained a solution of the transient energy equation applied to a step temperature change for the sphere. Their solution is in terms of an infinite series. The solution of Eq. (1.76) with the history integral is the general solution that may be applied in all transient processes. Using the same analytical method, Sazhin et al. (2001) derived an identical equation and confirmed the presence of the history term in the transient energy exchange equation. Also, Coimbra et al. (1998) and Coimbra and Rangel (2000) used the method of fractional calculus to derive the same transient equation for spheres and included simple radiation effects.

Regarding the effects of the history term on the heat transfer from a solid sphere, Gay and Michaelides (2003) calculated numerically the contribution of this term on the transient rate of energy exchange at creeping flow conditions and concluded that the contribution of the history term depends on the thermal Stokes number of the particle, St_{th}, which is defined as the ratio of the thermal diffusion timescale of the particle, $\tau_{th} = 4\alpha^2 \rho_f c_f/k_f$, to the characteristic timescale of the fluid $\tau_f = 2\alpha/U_{ch}$. Depending on the properties of the fluid and the sphere and on the relevant timescales, the heat transfer enhancement from a single sphere, due to the history term, is in the range of 0–50%.

It must be pointed out that the added mass term in the equation of motion derives from the pressure gradient term in the governing equation (1.17). Since the governing equation for the energy transfer does not have a corresponding term equivalent to ∇P, there is no term corresponding to the added mass term in the solution of the transient energy equation. The absence of a corresponding term for the added mass is the main difference in the functional forms of the equations of motion and temperature variation under the creeping flow conditions as well as at finite Re_s. For this reason, there is not a strict similarity between the transient equations of motion and the energy for particles, drops, or bubbles, although there

---

[3] For incompressible substances such as solids and liquids, the specific heats at constant pressure and constant volume are equal and denoted simply by the symbol c, $c_p = c_v = c$.

have been observed several analogous results and analogous behavior in the transient momentum and heat exchange processes for viscous and solid spheres.

1.4.4.2   At Finite but Small Peclet Numbers

Feng and Michaelides (1988) performed a study on the energy equation of a particle with arbitrary motion and in an arbitrary temperature field at finite but small Peclet numbers ($Pe_s < 1$). This study is analogous to the one performed by Lovalenti and Brady (1993a) for the equation of motion. A transient expression for the heat transfer from a solid particle of arbitrary shape, undergoing arbitrary motion, with a velocity given by the vector $\mathbf{u}^s$ was derived. At short dimensionless times, which are less than $O(Pe_s^{-2})$, the advection effects are insignificant, and the conduction solution applies in this case as well. At time scales that are higher than $O(Pe_s^{-2})$, advection becomes significant, and the total dimensionless heat transfer is expressed in terms of Pe_s and the time from the inception of the transient motion and heat transfer. In the case of a solid sphere undergoing a step temperature change, the transient heat transfer coefficient may be derived analytically and is given by the following expressions:

$$
\begin{aligned}
Nu &= 2 + \frac{Pe_s}{2} + \frac{2}{\sqrt{\pi t*}} \quad \text{for } t* < Pe_s^{-2} \\[2mm]
Nu &= 2 + \frac{Pe_s}{2} + 2\left[\frac{\exp\left(-\frac{Pe_s^2}{16} t*\right)}{\sqrt{\pi t*}} + \frac{Pe_s}{2} erf\left(\sqrt{\frac{Pe_s^2}{16} t*}\right)\right] \quad \text{for } t* > Pe_s^{-2}.
\end{aligned}
\tag{1.77}
$$

The quantity $t* = k_f t/4\alpha^2 \rho_f c_f$, which appears in the last two expressions, is the dimensionless time, based on the thermal timescale, $\tau_{th} = k_f/4\alpha^2 \rho_f c_f$. It is apparent in Eq. (1.77) that at short times, the rate of decay of the transient terms follows the typical $t^{-1/2}$ relationship of diffusion processes. At longer times ($t* > Pe_s^{-2}$), the advection effects become dominant, temperature gradients are swept by advection, and the transients decay significantly faster. At the longer times, the temperature gradients that form around the sphere have been advected to distances further than the so-called Oseen distance for the energy-transport process, which is of the order of αPe_s^{-1}.

Pozrikidis (1997) performed an analytical study to determine the transient heat and mass transfer from a suspended particle of arbitrary shape at low Peclet numbers. He used the method of matched asymptotic expansions and derived analytical asymptotic expressions for the energy transport from a sphere in a fluid undergoing a step temperature change and a sphere in a time-periodic flow. His results are expressed as the fractional increase of the rate of heat transfer from the case of pure conduction, $(Q(t) - Q_0)/Q_0$, where Q_0 is the result for pure conduction, and may be summarized as follows:

For uniform flow : $(Q(t) - Q_0)/Q_0 = 1/4\,Nu_0 Pe_s$
For simple shear flow : $(Q(t) - Q_0)/Q_0 = 0.1285\,Nu_0 Pe_\gamma^{1/2}$
For 2D straining flow : $(Q(t) - Q_0)/Q_0 = 0.68\,Nu_0 Pe_\gamma^{1/2}$ (1.78)
For axisymmetric straining flow : $(Q(t) - Q_0)/Q_0 = 1.0\,Nu_0 Pe_\gamma^{1/2}$

In the last four equations, Pe_s is the instantaneous Peclet number, $Pe_s(t)$, and Pe_γ is the instantaneous Peclet number based on the shear rate of the flow field, γ. These expressions confirm the fact that the velocity field developed around the particle influences significantly the instantaneous rate of heat transfer.

1.4.4.3   At Higher Peclet Numbers

The non-linear advection effects become dominant at $Pe_s \gg 1$, and for this reason, analytical or asymptotic solutions to the governing energy equation may not be obtained. Several results are known from the numerical studies that have been performed on this subject. Among these, Abramzon and Elata (1984) performed a numerical study and derived results for the transient Nusselt number, $Nu(t)$, for spheres undergoing a step temperature change, at high Pe_s, under a Stokesian velocity field, which necessarily implies that $Re_s < 1$. Feng and Michaelides (2000b) determined both the velocity and the temperature fields around the sphere in a uniform flow field, undergoing a step temperature change, in the ranges $0 < Re_s < 2{,}000$ and $0 < Pe_s < 2{,}000$. Among the conclusions of this study are:

(a) The short-time thermal behavior of the sphere at $t^* < O(Pe_s^{-1})$ is dominated by conduction effects. The Nusselt number is almost independent of the value of Re_s and may be adequately described by the short-term solution of Eq. (1.77).
(b) For $t* > O(Pe_s^{-1})$ and $Re_s > 2$, advection effects dominate. The value of instantaneous Re_s affects significantly the transient as well as the steady value of $Nu(t)$.
(c) The duration of the transients decreases dramatically with the increase of Pe_s or of Re_s.

Figure 1.4, which depicts the instantaneous Nusselt number versus the dimensionless time for several values of Re_s, shows typical results of this study.

Balachandar and Ha (2001) performed another, more general numerical study on the transient heat transfer from a sphere in a uniform flow field subjected to one of the following temperature changes: (a) a sudden step change of its temperature; (b) a step change of the temperature of the surrounding fluid; and (c) an oscillatory temperature change. It is apparent in this study that there are qualitative as well as quantitative differences in the behavior of the sphere under the three imposed temperature changes. For a step change of the fluid temperature, Balachandar and Ha (2001) concluded that the evolution of the dimensionless temperature of the

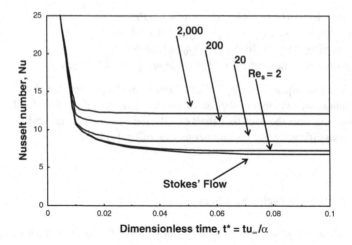

Fig. 1.4 Instantaneous Nusselt number for several values of Re_s at $Pe_s = 100$

sphere may be expressed by an exponential function, which is analogous to the solution obtained for the conduction problem:

$$T_s^* = 1 - \exp(-t/\tau_{eff}), \tag{1.79}$$

where τ_{eff} depends on the properties of the sphere and the fluid as well as the instantaneous Peclet number. Balachandar and Ha (2001) also obtained the behavior of the kernel of the history term. They showed that, at high Re_s, this kernel decays significantly faster than the $t^{-1/2}$ term, which characterizes the creeping flow decay of Eq. (1.76).

1.4.5 Turbulence Effects

Free-stream turbulence agitates the flow field; re-distributes particles, drops, and bubbles; and results in an increase of the heat transfer coefficients. Clift et al. (1978) compiled several sets of experimental data for particles and suggested the following empirical dependence of the Nusselt number on turbulence:

$$\frac{Nu}{Nu_0} = 1.0 + 4.8 \times 10^{-4} \frac{I_r}{I_{rc}} Re_s^{0.57}, \tag{1.80}$$

where Nu_0 is the Nusselt number of the particle in the absence of free-stream turbulence; $I_r = u_{rms}/|u_f - v_s|$ is the relative turbulence intensity with u_{rms} being the r.m.s of the velocity fluctuations; and I_{rc} is a critical relative turbulence intensity, which is defined as follows:

$$I_{rc} = [5.477 - \log_{10}(Re_s)]/15.8 \quad \text{for } I_r \leq 0.15$$
$$I_{rc} = [3.71 - \log_{10}(Re_s)]/1.75 \quad \text{for } I_r > 0.15. \tag{1.81}$$

A more recent study on the effect of free-stream turbulence by Yearling and Gould (1995) conducted with evaporating droplets of water, ethanol, and methanol generated the following empirical correlation for Nu:

$$\frac{Nu}{Nu_0} = 1 + 3.4I_r^{0.843}. \tag{1.82}$$

Yearling and Gould (1995) considered evaporating droplets, and hence, Nu_0 is the Nusselt number in the presence of blowing effects, which is given by equation (1.73). The film properties must be used for the properties of the droplets and the carrier fluid in the last equation.

1.4.6   Heat Transfer from Irregularly Shaped Particles

While most of the small bubbles and drops have spherical shapes, the majority of the solid particles have irregular shapes. Unlike the drag coefficients for irregular particles, where generalized expressions for C_D have been obtained from experimental data, there are no such generalized correlations for the heat transfer coefficients. In most practical situations, the heat transfer from irregular particles is approximated with the heat transfer from a sphere with the same area-equivalent diameter, d_A.

The only analytical model on the heat transfer from non-spherical particles was developed by Douglas and Churchill (1956): At finite Peclet numbers, where advection is important, a wake is formed behind the particle, and the flow separates from its surface. The heat transfer process from any particle, according to this model, is considered as the sum of two contributions that are produced from the rates of heat transfer forward of the separation line and aft of the separation line. It has been observed experimentally that, for a wide range of shapes and flows, the separation occurs approximately at the maximum perimeter normal to the flow. Douglas and Churchill (1956) obtained the following correlation for the Nusselt number at the aft portion of a three-dimensional, irregular object:

$$Nu_{aft} - Nu_{0\,aft}/2 = 0.056(Re_{aft})^{0.71}Pr^{1/3}, \tag{1.83}$$

where the Reynolds number Re_{aft} is defined in terms of a length equal to the ratio of the aft surface area to the maximum perimeter of the particle normal to the flow direction. The forward part of the rate of heat transfer is added to the heat transfer from the aft of the particle. For the forward part, Douglas and Churchill (1956) assumed that Nu has the same dependence on Re_s as the solid spheres, that is, Nu is proportional to $Re_s^{0.5}$. They adjusted the constants to fit the experimental data and obtained the following approximate correlation for the Nusselt number for the heat transfer from irregular particles:

$$Nu = 1 + 0.056\left(\frac{A_{\text{aft}}}{A}Re_s\right)^{0.71} Pr^{1/3} + 0.62\left(1 - \frac{A_{\text{aft}}}{A}Re_s\right)^{0.5} Pr^{1/3}. \qquad (1.84)$$

The area, A_{aft}, in the last equations is defined as the surface area of the irregular particle that is downstream the maximum perimeter, which is normal to the flow. All the dimensionless numbers are defined in terms of a characteristic length, L_{ch}, which is equal to the ratio of the total surface area of the particle, divided by the maximum perimeter projected on a plane normal to the flow. Hence, the geometric characteristics of the particle are implicit in the expressions for Nu and Re_s.

For particles with fore and aft symmetry, where $A_{\text{aft}}/A = 0.5$, the last equation yields

$$Nu = 1 + 0.034(Re_s)^{0.71}Pr^{1/3} + 0.44(Re_s)^{0.5}Pr^{1/3}. \qquad (1.85)$$

It must be noted that these correlations are based on an approximate decomposition of the heat transfer process, which is based on the shape of the particle and may only be considered as a first approximation to the heat or mass transfer of irregular particles. Numerical results for specific shapes would yield more accurate heat transfer coefficients.

A third approximation method for irregular particles is to consider an upper and a lower bound for the Nusselt number and to take the average value of the two extremes. The upper bound is the Nusselt number of the sphere that circumscribes the object and the lower bound is the Nusselt number of the volume-equivalent sphere, which is the Nusselt number based on the diameter d_V.

1.4.7 Rarefaction and Interface Temperature Discontinuity Effects

All the expressions for the equation of motion and the energy equation of particles that have been presented so far were derived assuming that the velocity and temperature functions are continuous at the fluid–solid interface. This is commonly called the *no-slip* condition at the interface. However, it has been experimentally observed that, when the size of particles is of the same order of magnitude as the mean free path of the fluid, there is significant slip/discontinuity at the interface, both for the velocity and for the temperature. The slip affects significantly both the drag and heat transfer coefficients. Because the slip at the interface modulates the velocity and temperature gradients, any velocity and temperature discontinuities/

slip at the fluid-sphere interface would reduce the hydrodynamic force and the rate of heat transfer.

Interface slip is of relevance to fluids containing nanoparticles, the so-called *nanofluids*, which are examined in detail in Chap. 4. Several applications also encounter fluids with nano-bubbles and nano-drops. The particle size in nanofluids is comparable to the mean free path of the fluid, and the corresponding Knudsen number, *Kn*, is of the order of one. Feng and Michaelides (2012) performed an asymptotic analysis of a sphere, which may be viscous or inviscid, and derived the following expression for *Nu* in terms of a thermal slip parameter, ξ:

$$Nu = \frac{2}{1+\xi} + \frac{1}{2(1+\xi)^2}Pe_s + \frac{F}{4(1+\xi)^2}Pe_s^2 \ln(Pe_s) + \frac{1}{2(1+\xi)^2}$$

$$\times \left[\frac{-156 + 148F - 152\xi + 341\xi F + 129F^2 + 528\xi F^2}{960(1+2\xi)}\right]Pe_s^2$$

$$+ \frac{1}{4(1+\xi)^2}(\ln\gamma_{Eu} - \ln 2 + 2\xi\ln\gamma_{Eu})FPe_s^2 + \frac{F}{16(1+\xi)^3}Pe_s^3$$

$$\times \ln(Pe_s), \tag{1.86}$$

where $\ln\gamma_{Eu}$ is the natural logarithm of the Euler parameter, which is equal to 0.577215; *F* is the dimensionless drag force on the sphere, which depends on the velocity slip as well as on the ratio of viscosities, λ, in the case of viscous spheres (Feng and Michaelides 2012; Feng et al. 2012); and the dimensionless thermal slip parameter is given by the expression:

$$\xi = \frac{4(2-\zeta)k}{(k-1)\zeta}\frac{Kn}{Pr}, \tag{1.87}$$

where *k* is the ratio of the specific heats of the fluid and ζ is the *accommodation coefficient*, a molecular property of gases and liquids (Mikami et al. 1966). It must be noted that, because the molecular free path of liquids is significantly smaller than that of gases, the slip effects are more significant in gas-particle processes than in liquid-particle processes.

1.4.8 Radiation Effects

Thermal radiation is the mode of heat transfer through which electromagnetic energy is continuously emitted by a body. The energy emission applies to the system under observation as well as all the other systems that exist in its surroundings. For a sphere with surface temperature T_s, the emitted radiation power is equal to

$$\dot{Q}_{rad}^{em} = 4\sigma\varepsilon\pi\alpha^2 T_s^4, \tag{1.88}$$

where σ is the Stefan-Boltzmann constant, $\sigma = 5.669 \times 10^{-8}$ W/m^{2}K^4, and ε is the emissivity of the sphere, which depends on the material of its surface. The emissivity of a black body is equal to one. Similarly the sphere absorbs heat from all the objects in its surroundings. In the simple case where the sphere is enclosed by a single medium of temperature T_∞, the sphere absorbs thermal radiation equal to:

$$\dot{Q}_{rad}^{ab} = 4\sigma\alpha_s\pi\alpha^2 T_\infty{}^4, \tag{1.89}$$

where α_s is the absorptivity of the sphere. In general, the absorptivity and the emissivity of a material are functions of its temperature and are equal in magnitude: $\alpha_s(T_s) = \varepsilon(T_s)$.

The net rate of energy that enters the sphere as a result of thermal radiation is equal to the difference of the above two equations. In analogy with the convective heat transfer, it is often expressed in terms of a thermal radiation coefficient, h_{rad}:

$$\dot{Q}_{rad}^{ab} - \dot{Q}_{rad}^{em} = 4\sigma\varepsilon\pi\alpha^2 (T_\infty{}^4 - T_s{}^4) = 4\pi\alpha^2 h_{rad}(T_\infty - T_s). \tag{1.90}$$

It is apparent that the thermal radiation coefficient is a strong function of temperature. When the temperature difference of the sphere and its surroundings is sufficiently high, the net radiation power is significant and may actually surpass the energy exchange due to the other two modes of heat transfer, conduction and convection. The total power exchange between the sphere and its surroundings is the sum of convection and radiation:

$$\dot{Q}_T = 4\pi\alpha^2 (h_{rad} + h)(T_\infty - T_s). \tag{1.91}$$

It must be noted that, in the case of radiation, one has to account not only for the carrier gas, but also for all the boundaries and all other objects that may radiate energy to the sphere. All objects exchange radiation with any other objects or surfaces they "see," or have a clear path for photon exchange. In the general case, when the object under study is surrounded by several other objects, including other similar objects in the same carrier fluid, the determination of the net thermal radiation or the thermal radiation coefficient must be carried out by carefully evaluating the effects of all the surfaces in the surroundings. This may become a challenging task in complex engineering systems with several boundaries and particles, such as fluidized bed reactors. A more extensive description of the processes and the methods used for the determination of radiative heat transfer may be found in specialized treatises on radiation, such as the one by Siegel and Howel (1981).

1.4.9 Temperature Measurements

Radiation from a particle has important ramifications on temperature measurements. Let us consider a thermometer immersed inside a fluid stream of true temperature

Fig. 1.5 Temperature measurement in a gas surrounded by radiating boundaries

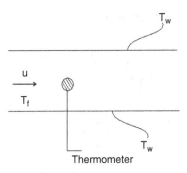

Thermometer

T_f and velocity u, as shown in Fig. 1.5. The flow is inside a long conduit whose walls are at a different temperature T_w. Since the thermometer exchanges heat with the fluid by convection and with the walls by radiation, at steady state, its temperature, T_{th}, is defined by the heat transfer equilibrium between convection and radiation. In the simplest case of black-body radiation and a non-absorbing gas, this equilibrium yields the following expression for the temperature of the thermometer:

$$h_c(T_f - T_{th}) = h_{rad}(T_{th} - T_w) \Rightarrow T_{th} = \frac{h_c T_f + h_{rad} T_w}{h_c + h_{rad}}, \qquad (1.92)$$

where h_c and h_{rad} are the convective and radiative heat transfer coefficients. The last equation implies that the thermometer will measure approximately the true temperature of the gas, T_f, only when $h_c \gg h_{rad}$. On the contrary, the thermometer will measure the temperature of the walls, T_w, when $h_c \ll h_{rad}$, and an intermediate temperature, which is not the true temperature of the fluid, T_f, under any other condition. Therefore, any value of temperature measured by thermometers exposed to surrounding radiative heat transfer must be corrected by this expression.

In the case of transient temperature measurements, the thermal timescale of the thermometer must be also taken under consideration. To obtain any meaningful measurements, the thermometer must be in thermodynamic equilibrium with the object, whose temperature is measured. This implies that the characteristic time of response of the thermometer must be significantly smaller than the characteristic time of the temperature variation of the object whose temperature is measured. If this condition is not satisfied, the measurements obtained are not meaningful. For thermocouples, which may be approximated as small solid spheres, their characteristic time is the thermal timescale of the almost-spherical tip, τ_{th}. For any meaningful transient temperature measurements, the characteristic time of temperature fluctuations, τ_{char}, must be significantly greater than the characteristic time of the instrument τ_{th}, that is, $\tau_{char} \gg \tau_{th}$. For this reason, very small thermocouples, which have very low values of τ_{th}, are the instruments of choice for transient temperature measurements.

Bibliography

Abramzon B, Elata C (1984) Heat transfer from a single sphere in stokes flow. Int J Heat Mass Transf 27:687–695

Abramzon B, Sirignano WA (1989) Droplet vaporization for spray combustion calculations. Int J Heat Mass Transf 32:1605–1618

Achenbach E (1974) Vortex shedding from spheres. J Fluid Mech 62:209–221

Acrivos A (1980) A note on the rate of heat or mass transfer from a small particle freely suspended in linear shear field. J Fluid Mech 98:299–304

Acrivos A, Goddard JD (1965) Asymptotic expansions for laminar convection heat and mass transfer. J Fluid Mech 23:273–291

Acrivos A, Taylor TE (1962) Heat and mass transfer from single spheres in stokes flow. Phys Fluids 5:387–394

Al-taweel AM, Carley JF (1971) Dynamics of single spheres in pulsated flowing liquids: Part I. Experimental methods and results. AIChE Symp Ser 67(116):114–123

Balachandar S, Ha MY (2001) Unsteady heat transfer from a sphere in a uniform cross-flow. Phys Fluids 13(12):3714–3728

Basset AB (1888) Treatise on hydrodynamics. Bell, London

Boussinesq VJ (1885) Sur la Resistance qu' Oppose un Liquide Indéfini en Repos .... Comptes Rendu Acad Sci Paris 100:935–937

Brunn PO (1982) Heat or mass transfer from single spheres in a low Reynolds number flow. Int J Eng Sci 20:817–822

Carslaw HS, Jaeger JC (1947) Conduction of heat in solids. Oxford University Press, Oxford

Chhabra RP, Singh T, Nandrajog S (1995) Drag on chains and agglomerates of spheres in viscous Newtonian and power law fluids. Can J Chem Eng 73:566–571

Chiang CH, Raju MS, Sirignano WA (1992) Numerical analysis of a convecting, vaporizing fuel droplet with variable properties. Int J Heat Mass Transf 35:1307–1327

Clift KA, Lever DA (1985) Isothermal flow past a blowing sphere. Int J Numer Methods Fluids 5:709–715

Clift R, Grace JR, Weber ME (1978) Bubbles, drops and particles. Academic, New York

Coimbra CFM, Rangel RH (2000) Unsteady heat transfer in the harmonic heating of a dilute suspension of small particles. Int J Heat Mass Transf 43:3305–3316

Coimbra CFM, Edwards DK, Rangel RH (1998) Heat transfer in a homogenous suspension including radiation and history effects. J Thermophys Heat Transf 12:304–312

Crowe CT, Sommerfeld M, Tsuji Y (1998) Multiphase flows with droplets and particles. CRC, Boca Raton, FL

Dandy DS, Dwyer HA (1990) A sphere in shear flow at finite Reynolds number: effect of particle lift, drag and heat transfer. J Fluid Mech 218:381–412

Douglas WJM, Churchill SW (1956) Heat and mass transfer correlations for irregular particles. Chem Eng Symp Ser 52(18):23–28

Faxen H (1922) Der Widerstand gegen die Bewegung einer starren Kugel in einer zum den Flussigkeit, die zwischen zwei parallelen Ebenen Winden eingeschlossen ist. Ann Phys 68:89–119

Feng Z-G, Michaelides EE (1998) Transient heat transfer from a particle with arbitrary shape and motion. J Heat Transf 120:674–681

Feng Z-G, Michaelides EE (2000a) A numerical study on the transient heat transfer from a sphere at high Reynolds and Peclet numbers. Int J Heat Mass Transf 43:219–229

Feng Z-G, Michaelides EE (2000b) Mass and heat transfer from fluid spheres at low Reynolds numbers. Powder Technol 112:63–69

Feng Z-G, Michaelides EE (2001a) Drag coefficients of viscous spheres at intermediate and high Reynolds numbers. J Fluids Eng 123:841–849

Feng Z-G, Michaelides EE (2001b) Heat and mass transfer coefficients of viscous spheres. Int J Heat Mass Transf 44:4445–4454

Feng Z-G, Michaelides EE (2002) Inter-particle forces and lift on a particle attached to a solid boundary in suspension flow. Phys Fluids 14:49–60

Feng Z-G, Michaelides EE (2012) Heat transfer from a nano-sphere with temperature and velocity discontinuities at the interface. Int J Heat Mass Transf 55(23–24):6491–6498

Feng Z-G, Michaelides EE, Mao S-L (2012) On the drag force of a viscous sphere with interfacial slip at small but finite Reynolds numbers. Fluid Dyn Res 44:025502. doi:10.1088/0169-5983/44/2/025502

Fourier J (1822) Theorie Analytique de la Chaleur. Paris

Galindo V, Gerbeth G (1993) A note on the force on an accelerating spherical drop at low Reynolds numbers. Phys Fluids 5:3290–3292

Gay M, Michaelides EE (2003) Effect of the history term on the transient energy equation of a sphere. Int J Heat Mass Transf 46:1575–1586

Hadamard JS (1911) Mouvement Permanent Lent d' une Sphere Liquide et Visqueuse dans un Liquide Visqueux. Compte-Rendus de' l' Acad des Sci Paris 152:1735–1738

Haider AM, Levenspiel O (1989) Drag coefficient and terminal velocity of spherical and nonspherical particles. Powder Technol 58:63–70

Harper JF (1972) The motion of bubbles and drops through liquids. Adv Appl Mech 12:59–129

Hartman M, Yates JG (1993) Free-fall of solid particles through fluids. Collect Czech Chem Commun 58:961–974

Karanfilian SK, Kotas TJ (1978) Drag on a sphere in unsteady motion in a liquid at rest. J Fluid Mech 87:85–96

Kim S, Karila SJ (1991) Microhydrodynamics: principles and selected applications. Butterworth-Heineman, Boston, MA

Kurose R, Makino H, Komori S, Nakamura M, Akamatsu F, Katsuki M (2003) Effects of outflow from surface of sphere on drag, shear lift and scalar diffusion. Phys Fluids 15:2338–2351

Lasso IA, Weidman PD (1986) Stokes drag on hollow cylinders and conglomerates. Phys Fluids 29(12):3921–3934

Leal LG (1992) Laminar flow and convective transport processes. Butterworth-Heinemann, Boston, MA

Lerner SL, Homan HS, Sirignano WA (1980) Multicomponent droplet vaporization at high Reynolds numbers-size, composition and trajectory histories. In: Proceedings of AIChE annual meeting, Chicago

Levich VG (1962) Physicochemical hydrodynamics. Prentice-Hall, Englewood Cliffs, NJ

Ling Y, Haselbacher A, Balachandar S (2011a) Importance of unsteady contributions to force and heating for particles in compressible flows. Part 1: Modeling and analysis for shock–particle interaction. Int J Multiphase Flow 37:1026–1044

Ling Y, Haselbacher A, Balachandar S (2011b) Importance of unsteady contributions to force and heating for particles in compressible flows. Part 2: Application to particle dispersal by blast waves. Int J Multiphase Flow 37:1013–1025

Loth E (2000) Numerical approaches for the motion of dispersed particles, droplets, or bubbles. Prog Energy Combust Sci 26:161–223

Lovalenti PM, Brady JF (1993a) The hydrodynamic force on a rigid particle undergoing arbitrary time-dependent motion at small Reynolds numbers. J Fluid Mech 256:561–601

Lovalenti PM, Brady JF (1993b) The force on a bubble, drop or particle in arbitrary time-dependent motion at small Reynolds numbers. Phys Fluids 5:2104–2116

Madhav GV, Chhabra RP (1995) Drag on non-spherical particles in viscous fluids. Int J Miner Process 43:15–29

Magnus G (1861) A note on the rotary motion of the liquid jet. Annalen der Physik und Chemie 63:363–365

Maxey MR, Riley JJ (1983) Equation of motion of a small rigid sphere in a non-uniform flow. Phys Fluids 26:883–889

Maxworthy T (1969) Experiments on the flow around a sphere at high Reynolds numbers. J Appl Mech 91:598–607

McLaughlin JB (1991) Inertial migration of a small sphere in linear shear flows. J Fluid Mech 224:261–74

Mei R (1992) An approximate expression of the shear lift on a spherical particle at finite Reynolds numbers. Int J Multiphase Flow 18:145–160

Mei R, Adrian RJ (1992) Flow past a sphere with an oscillation in the free-stream and unsteady drag at finite Reynolds number. J Fluid Mech 237:323–341

Mei R, Lawrence CJ, Adrian RJ (1991) Unsteady drag on a sphere at finite Reynolds number with small fluctuations in the free-stream velocity. J Fluid Mech 233:613–631

Michaelides EE (2003) Hydrodynamic force and heat/mass transfer from particles, bubbles and drops—The Freeman Scholar Lecture. J Fluids Eng 125:209–238

Michaelides EE (2006) Particles, bubbles and drops – their motion, heat and mass transfer. World Scientific, Hackensack, NJ

Michaelides EE, Feng Z-G (1994) Heat transfer from a rigid sphere in a non-uniform flow and temperature field. Int J Heat Mass Transf 37:2069–2076

Michaelides EE, Feng Z-G (1995) The equation of motion of a small viscous sphere in an unsteady flow with interface slip. Int J Multiphase Flow 21:315–321

Michaelides EE, Roig A (2011) A reinterpretation of the Odar and Hamilton data on the unsteady equation of motion of particles. AIChE J. doi:10.1002/aic.12498, 57, 2997-3002

Mikami H, Endo Y, Takashima Y (1966) Heat transfer from a sphere to rarefied gas mixtures. Int J Heat Mass Transf 9:1435–1448

Odar F, Hamilton WS (1964) Forces on a sphere accelerating in a viscous fluid. J Fluid Mech 18:302–303

Oesterle B, Bui Dinh A (1998) Experiments on the lift of a spinning sphere in the range of intermediate Reynolds numbers. Exp Fluids 25:16–22

Oliver DL, Chung JN (1987) Flow about a fluid sphere at low to moderate Reynolds numbers. J Fluid Mech 177:1–18

Parmar M, Haselbacher A, Balachandar S (2011) Generalized Basset-Boussinesq-Oseen equation for unsteady forces on a sphere in a compressible flow. Phys Rev Lett 106:084501

Pettyjohn ES, Christiansen ER (1948) Effect of particle shape on free-settling rates of isometric particles. Chem Eng Prog 44:157–172

Pozrikidis C (1997) Unsteady heat or mass transport from a suspended particle at low Peclet numbers. J Fluid Mech 289:652–688

Proudman I, Pearson JRA (1956) Expansions at small Reynolds numbers for the flow past a sphere and a circular cylinder. J Fluid Mech 2:237–262

Ranz WE, Marshall WR (1952) Evaporation from drops. Chem Eng Prog 48:141–146

Rybczynski W (1911) On the translatory motion of a fluid sphere in a viscous medium. Bull Acad Sci Krakow Ser A 40–46

Saffman PG (1965) The lift on a small sphere in a slow shear flow. J Fluid Mech 22:385–398

Saffman PG (1968) The lift on a small sphere in a slow shear flow—corrigendum. J Fluid Mech 31:624–625

Sano T (1981) Unsteady flow past a sphere at low Reynolds number. J Fluid Mech 112:433–441

Sazhin SS, Goldstein VA, Heikal MR (2001) A transient formulation of Newton's cooling law for spherical bodies. J Heat Transf 123:63–64

Schiller L, Nauman A (1933) Uber die grundlegende Berechnung bei der Schwekraftaufbereitung. Ver Deutch Ing 44:318–320

Seeley LE, Hummel RL, Smith JW (1975) Experimental velocity profiles in laminar flow around spheres at intermediate Reynolds numbers. J Fluid Mech 68:591–608

Siegel R, Howel JR (1981) Thermal radiation heat transfer. McGraw-Hill, New York

Sirignano WA (1999) Fluid dynamics and transport of droplets and sprays. Cambridge University Press, Cambridge

Soo SL (1990) Multiphase fluid dynamics. Science, Beijing

Sridhar G, Katz J (1995) Drag and lift forces on microscopic bubbles entrained by a vortex. Phys Fluids 7(2):389–399

Tanaka T, Yamagata K, Tsuji Y (1990) Experiment on fluid forces on a rotating sphere and a spheroid. In: Proceedings of 2nd KSME-JSME fluids engineering conference, vol 1, pp 266–378

Taneda S (1956) Experimental investigation of the wake behind a sphere at low Reynolds numbers. J Phys Soc Jpn 11(10):1104–1108

Tomiyama A, Tamai H, Zun I, Hosokawa S (1999) Transverse migration of single bubbles in simple shear flows. In: Proceedings of 2nd international symposium on two-phase flow modeling and experimentation, vol 2, pp 941–948

Tran-Cong S, Gay M, Michaelides EE (2004) Drag coefficients of irregularly shaped particles. Powder Technol 139:21–32

Tsuji Y, Morikawa Y, Mizuno O (1985) Experimental measurements of the magnus force on a rotating sphere at low Reynolds numbers bubble in an axisymmetric shear flow. J Fluids Eng 107:484–498

Vojir DJ, Michaelides EE (1994) The effect of the history term on the motion of rigid spheres in a viscous fluid. Int J Multiphase Flow 20:547–556

Wadell H (1933) Sphericity and roundness of rock particles. J Geol 41:310–331

Whitaker S (1972) Forced convection heat transfer correlations for flow in pipes past flat plates, single cylinders, single spheres, and for flow in packed beds and tubes bundles. AIChE J 18:361–371

Winnikow S, Chao BT (1966) Droplet motion in purified systems. Phys Fluids 9:50–61

Yearling PR, Gould RD (1995) Convective heat and mass transfer from a single evaporating water, ethanol and methanol droplet. Proc ASME-FED 233:33–39

Yuen MC, Chen LW (1976) On drag of evaporating droplets. Combust Sci Technol 14:147–154

Chapter 2
Numerical Modeling and Simulations

Keywords Model attributes • Particle-fluid coupling • Two-fluid • Point-source • Direct simulation • Collisions • Coalescence • Boundaries

2.1 Multiphase and Particulate Modeling

The design of processes and equipment was traditionally accomplished in the past by experimentation, construction of prototypes, and the building of pilot plants. These methods have been time-consuming and labor-intensive and, in the beginning of the twenty-first century, have been proven to be very expensive. In the last two decades, modeling and computer simulations are increasingly used for the design of equipment and processes. The main advantage of computer simulations is that they require significantly less time and resources than the building, testing, and optimization of prototypes and pilot plants. The main disadvantage of simulations is that, oftentimes, the modeling does not accurately describe the actual engineering system to be built, and some or all the testing results suffer from inaccuracies. This appears to be a temporary drawback to the simulation methods because it is due to the fact that numerical simulations are recently developed methods in science and engineering. The art and science of simulations are in the early stages of development, and still, there is not a great deal of accumulated expertise on this subject. Such inaccuracies are related to lack of modeling knowledge or lack of understanding of physical phenomena associated with the systems or processes that are modeled. With the continuing research and the advancement of our knowledge of phenomena and systems to be simulated, the modeling techniques become better and the accuracy of the simulations continuously improves. In addition, the continuous improvements in computer algorithms and the increasing availability of more powerful computers and processing time will assist significantly the accuracy and reliability of simulations in the future.

E.E. (Stathis) Michaelides, *Heat and Mass Transfer*
in Particulate Suspensions, SpringerBriefs in Applied Sciences and Technology,
DOI 10.1007/978-1-4614-5854-8_2, © Springer Science+Business Media New York 2013

The numerical simulation of systems and processes involves two stages:

(a) The modeling of the system or process by a set of equations, which are the governing and closure equations. The set ideally includes all the salient parts and features of the system or process and their interactions.
(b) The numerical implementation to obtain the solution of the governing equations. This includes the discretization of the set of equations and the numerical method to be followed in order to derive the solution.

From the beginning, it must be noted that a numerical model is a mathematical idealization of the system and not an exact replica of the system. The modeling will not replicate faithfully all the features of the system. However, a useful and "successful" model will faithfully reproduce the most important features, which are of interest to the modeler. The basic premise of the model is to provide reasonably accurate answers for the behavior of the system under different conditions. Therefore, the modeling process has to start with the inquiries that need to be answered for the modeled system. Some of these inquiries that are pertinent to particulate heat transfer systems are:

1. Is there an interest in the transient behavior of the system/process or is steady-state representation sufficient?
2. Is there an interest in the inhomogeneities of the distribution of particles or a space-averaged description is sufficient?
3. Is the system diluted—are the particles sparsely distributed—for the interactions of the particles to be neglected?
4. Does the system generate turbulence? Are the turbulence effects important for the system or process?
5. Is there a need to describe the motion and heat transfer from individual particles or a general/average description is adequate?
6. Are electrical or magnetic effects important to be modeled?
7. Are other effects important and need to be modeled?

Answers to these questions always assist in the modeling of the particulate system and guide the modeling and simulation processes. For example, if the answer to question 4 is negative, the modeling may be simplified considerably by the choice of laminar governing equations for the momentum and heat transfer. If the answer to this question is positive and turbulence is generated in the system, the modeler must use one of the several available turbulence models in order to accurately describe the turbulence in the carrier fluid flow. At this stage the modeler will have to decide how to include the particle–turbulence interactions in the model as one-way interaction or two-way interaction. In the latter case, the following two interactions will have to be modeled:

(a) Particles are dispersed by the turbulence field, and their motion and energy transfer are affected by the turbulent eddies.
(b) Particles modulate the carrier fluid turbulence and by extent, the velocity field of the carrier fluid.

It is apparent that the inclusion of both effects will increase the complexity of the model and the computational resources it requires. The advantage from the increased complexity is that the results of the model will be more accurate.

2.1.1   Desired Attributes of Models

A mathematical model is a set of equations that describes an actual system or process. Models are mathematical and often idealized representations of physical situations, not exact replicas of them. Therefore, a model may predict with a certain degree of accuracy the effects of some of the variables associated with a system or a process, but may not be capable to predict other parameters or, in extreme and undesirable cases, may even yield inaccurate predictions. For a model to be useful to engineers, whose function is to design or evaluate systems and processes, it is desirable to possess the following attributes:

1. Simplicity in the structure of its equations.
2. Use of well-defined parameters that are universally accepted.
3. Ease in its comprehension by the engineer or practitioner.
4. Accuracy, to the degree desired by the modeler, for the particular application and reliability for its predictions.
5. Computational robustness, simplicity in the coding of its elements, and relative simplicity of the required computational grid.
6. Generality in its applications. There is invariably a trade-off between generality and simplicity.
7. Clear path to the validation and verification of its results, by corroborating its predictions with parameters, which are readily and accurately measured in real applications of the model.
8. Completeness in computing all the needed parameters.
9. Correct asymptotic behavior when its results are extended to single-phase flows and energy transfer or to flows where the multiphase mixture is expected to behave as a single-phase fluid, such as very dilute bubbly systems or a dilute mixture of air with aerosol particles.
10. Agreement of the results with the empirical correlations that have been experimentally validated in the past and continue to be used in the design of processes and equipment.

Because some of the first applications of multiphase flows were in the design of industrial equipment related to boiling, condensing, refining, and heat exchange between fluids, the first simplified multiphase flow models were developed for pipe flows. The origin of these models is the area-averaged equations for a multiphase mixture (Delhaye 1981; Michaelides 2003) with several simplifications being made in order to facilitate closure and ease in computations. Delhaye (1981), Boure and Delhaye (1982), Wallis (1963), and Ishii (1975, 1990), among many others, have presented accounts of several of these models. In the sections that follow, the basic features and equations of models for the flow and energy transfer from particulate systems will be presented. Some of the numerical methods that are used to obtain the numerical solution of the governing equations will also be briefly exposed.

2.2 Classification of Particulate Flows

The first method of the classification of particulate flows is related to the volumetric composition of the particle–fluid mixture in *dilute* and *dense* flows. Dilute flows occur at volumetric concentrations, ϕ, less than 2%, and dense flows occur in general at $\phi > 6.5\%$. Interparticle collisions and particle interactions must always be taken into account in dense flows, while the two phenomena may be neglected in dilute flows without significant loss of accuracy. In the *intermediate* range $2\% < \phi < 6.5\%$, modeling of the collisions and interactions depends on the degree of the desired accuracy to be achieved by the numerical scheme to be employed.

The value of 6.5% concentration stems from the average interparticle distance: If the particulate phase is composed of spheres with a uniform diameter, $d = 2\alpha$, in the dense flow regime, the average distance between two spheres is less than one sphere diameter, when $\phi > 6.5\%$. This implies that hydrodynamic interactions between the spheres, and interparticle collisions, may not be neglected. In general, particles—this term includes both solid particles and liquid drops—interact with the carrier fluid by exchanging mass, momentum, and energy. The most important classification of particulate flows for modeling purposes is done according to the type and strength of these interactions, processes that are often called the *coupling*, between the carrier phase and the particulate phase. Accordingly, we have the following four classifications of particulate flows.

2.2.1 One-Way Coupling

This type of particulate flows assumes that the carrier flow field affects the motion of the particles through the hydrodynamic drag force, but the particles exert a negligible effect on the carrier fluid. One-way coupling implies that there is very low volumetric concentration of particles and these particles have a negligible effect on the flow of the carrier phase. The particles move independently within the carrier fluid and exchange momentum and energy with the fluid based on the drag and heat transfer expressions for single particles. Interactions between particles are neglected. This type of flow is modeled by solving separately for the velocity and temperature fields of the carrier fluid in the absence of particles and following individual particles in a Lagrangian frame of reference with origin the center of the particles. Oftentimes these simulations are called *Monte Carlo* (*MC*) simulations. Typical examples of one-way coupling are the dilute pneumatic conveying or drying of particles, and heat transfer with nanofluids at very low volumetric concentrations.

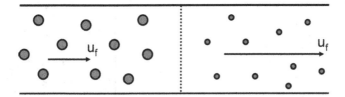

Fig. 2.1 Two-way coupling: The evaporation of drops in the two sections of this pipe causes the increase in the fluid volume and fluid velocity, which affects the transport velocity and heat transfer of the drops

2.2.2   Two-Way Coupling

In two-way coupling, the interactions of particles and carrier fluid are such that the effect of the interactions on the fluid may not be neglected, even though the volumetric flow of the particles is very low. A typical example of two-way coupling is the evaporation of drops in the last stages of boiling, which is depicted schematically in Fig. 2.1. Heat is transferred to the conduit, reaches the drops through the fluid, and causes their evaporation. Because the density of the drops is much larger than that of the produced vapor, the relatively small volume of the evaporating drops causes a significant increase in the volumetric flow of the vapor, which in a confined conduit accelerates significantly both fluid and drops.

Particles affect the carrier fluid in all chemical reactions where a vapor or gas is produced, including all combustion processes. Typically, in two-way coupling, the effects of the particles on the velocity and temperature of the carrier fluid are modeled as source terms in the mass, momentum, and energy equations. The motion of the individual particles or groups of particles may still be modeled in a Lagrangian way.

2.2.3   Three-Way Coupling

In addition to the effects of the carrier fluid on the particle motion and of particles on the fluid motion, in three-way coupling the hydrodynamic interactions between the particles, such as drafting in the wakes of preceding particles and lubrication effects, play an important role and are modeled. Also modeled are thermal interactions between particles, especially radiation heat transfer in combustors. Particles that produce heat, which is transferred to the surrounding fluid, induce a natural convection flow field around them. This field affects the velocity and temperature of the fluid and of other particles in the immediate vicinity. In the case of burning/reacting particles, the flow field emanating from the gases produced and the natural convection around particles is strong and must be modeled in a

Fig. 2.2 Schematic diagram
of a system, where modeling
with four-way coupling is
necessary

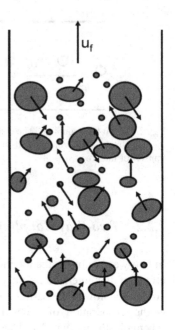

three-way modeling. Most applications of particulate systems with intermediate and dense concentrations, where combustion or evaporation occurs, are best to be modeled with a three-way coupling.

2.2.4 Four-Way Coupling

The modeling of interparticle collisions is the fourth element that characterizes the four-way coupling and is typically associated with dense flows. In a four-way coupling, the carrier fluid influences the motion and heat transfer from particles, and the particles affect the velocity and temperature of this fluid. In addition, the modeling takes into account the hydrodynamic interactions between the particles, particle–particle collisions, and particle–wall collisions. The flow and heat transfer in a fluidized bed reactor (FBR) as well as in most chemical reactors are typical examples, where four-way coupling is needed for the accurate modeling of the processes. Figure 2.2 is a schematic diagram of the velocities of particles in a FBR, where four-way coupling must be used for complete and accurate modeling. The velocities of several particles are shown by the representative vectors, $v(t)$. The velocity of the fluid is variable in space and time, $u_f = u_f(x_i,t)$, and equal to the particle velocity at each fluid–particle interface. It is apparent from this figure that hydrodynamic interactions occur between several particles as well as between groups of particles. The instantaneous velocities of the particles imply that several collisions are very likely to occur within a short time between particles as well as between several particles and the surrounding walls. The accurate and meaningful description of the mass, momentum, and energy transfer in this system necessitates the accurate modeling of the fluid–particle, particle–fluid, and particle–particle hydrodynamic and energetic interactions, as well as the modeling of collisions between particles and between particles and surrounding walls. The four-way coupling is typically carried out numerically by the Eulerian description of the two phases, which is sometimes referred to as the *two-fluid model*. Closure equations for all the interactions must be supplied to the two-fluid models. The closure equations may be derived from experimental data, analytical studies, or more detailed numerical studies.

2.3   Modeling of the Carrier Phase: Governing Equations

In modeling the carrier fluid, one must first consider whether or not the carrier fluid domain may be characterized as a continuum. Flows and energy exchange in nano-pores of membranes and chemical activation close to the surface of irregular catalyst particles are two examples where the molecular effects dominate and the carrier fluid in these regions may not be assumed to be a continuum. In most of the other applications of particulate heat transfer systems, the carrier fluid satisfies the continuum assumption. In this chapter, we will make use of the continuum assumption for the carrier fluid and will develop accordingly the governing equations for this fluid. Since most of the applications with particulates pertain to low Mach numbers, it will also be assumed that the carrier fluid flow is incompressible, that is, $\rho_f = $ const. Hence, the mass conservation equation for the fluid—or continuity equation—becomes

$$\frac{\partial u_i}{\partial x_i} = 0 \quad \text{or} \quad \nabla \cdot \vec{u} = 0. \tag{2.1}$$

The form of the momentum and energy equations of the carrier fluid depends on whether or not flow instabilities have developed. The fluid Reynolds number, Re_L, which is based on the characteristic dimension of the carrier flow, L, plays an important role in the characterization and modeling of the carrier fluid flow. The flow is *laminar*, and the fluid does not develop any instabilities if $Re_L < Re_{cr}$, where Re_{cr} is the critical Reynolds number at which instabilities are initiated in the carrier fluid. Re_{cr} depends strongly on the geometry of the system that is modeled and its boundaries. The flow is *turbulent* if $Re_L > Re_{tu}$, where Re_{tu} is another higher number, beyond which the flow is fully turbulent. When $Re_{cr} < Re_L < Re_{tu}$, the flow has developed instabilities but is not fully turbulent and is called *transitional*. Because of lack of knowledge and more accurate modeling methods, transitional flows are frequently modeled in the same way as turbulent flows.

2.3.1 Laminar Flow

When $Re_L < Re_{cr}$, any instabilities that may be developed in the flow decay fast and the flow is stable. The time-dependent momentum and energy equations of the carrier fluid for laminar flow may be written as follows:

$$\rho_f \left(\frac{\partial u_i}{\partial t} + u_j \frac{\partial u_i}{\partial x_j} \right) = \rho_f g_i - \frac{\partial P}{\partial x_i} + \mu_f \frac{\partial^2 u_i}{\partial x_j \partial x_j} \quad \text{or}$$

$$\rho_f \left(\frac{\partial \vec{u}}{\partial t} + \vec{u} \cdot \nabla \vec{u} \right) = \rho_f \vec{g} - \nabla P + \mu_f \nabla^2 \vec{u}, \tag{2.2}$$

and

$$\rho_f c_f \left(\frac{\partial T}{\partial t} + u_j \frac{\partial T}{\partial x_j} \right) = k_f \frac{\partial^2 T}{\partial x_j \partial x_j} \quad \text{or}$$

$$\rho_f c_f \left(\frac{\partial T}{\partial t} + \vec{u} \cdot \nabla T \right) = k_f \nabla^2 T. \tag{2.3}$$

For steady, laminar flow, the time-dependent terms vanish. In most applications, boundary layer approximations may be applied to appropriate regions of the system, e.g., near solid walls, which may simplify the stress tensor and the temperature gradients.

A special type of laminar flow is the *creeping flow* where $Re_L \ll 1$. Under creeping flow conditions, the nonlinear advection terms, which appear in the left-hand sides of the last two equations, are very small in comparison to the other terms and may be neglected. In creeping flow, both momentum and energy equations become linear and may be solved analytically. Such flows occur in micro- and nano-channels as well as in the vicinity of fine particles. The solution of the creeping flow equations around spheres has resulted in many of the analytical expressions for the transient hydrodynamic force and transient heat transfer coefficients of spheres, which were presented in Chap. 1.

2.3.2 Turbulent Flow

When $Re_L > Re_{tu}$, the flow instabilities have been developed and have been sufficiently amplified in the flow domain to be distinct and to have caused the formation of vortices, local jets, and other flow structures, which persist over large ranges of timescales and lengthscales. The flow structures that develop in turbulent flows are unsteady, three dimensional, and span several lengthscales, from the large eddy scales, L_{LE}, to the Kolmogorov microscale, $L_K = (\nu^3/\varepsilon)^{1/4}$. The latter is considered the smallest scale of turbulent eddies. At the higher lengthscales, the flow structures are almost deterministic, but at the lower lengthscales, they are essentially stochastic. The momentum and energy equations of the carrier fluid for turbulent flow are the same time-dependent expressions written for the laminar flow regime [Eqs. (2.2) and (2.3)].

A complete numerical description of a turbulent flow field, with accurate representation of the large and small eddies, is possible, but it is computationally very demanding because the ratio of the large to small eddies, L_{LE}/L_K, is proportional to $Re_L^{3/4}$. Simple calculations show that, for the simulation of water flow with velocity 1 m/s in a 5 cm pipe, $Re_L \approx 50,000$ and $L_{LE}/L_K \approx 3,344$. A direct computation for this rather simple flow system would require at least 8,000 grid points in each direction or about 500 billion grid points, a significant computational resource. For this reason, averaging methods have been developed for the description of turbulence. Among these, the *Reynolds decomposition* stipulates that the turbulent velocities and temperature may be decomposed to a time-averaged and a time-dependent component as follows:

$$u_i(t) = \bar{u}_i + u_i'(t) \tag{2.4}$$

and

$$T(t) = \bar{T}_i + T'(t), \tag{2.5}$$

where u' and T' are the *fluctuating velocity* and *fluctuating temperature*, respectively, and the bar represents time averaging. Both of these variables have zero mean, and the standard deviation of the former is the turbulence intensity (Hinze 1975). The usual procedure to solve the governing equations is to substitute the fluctuating velocity and temperature in Eqs. (2.2) and (2.3), time average the resulting expressions, and solve for the time-averaged velocity and temperature fields. The substitution of the decomposed velocity and temperature into the governing momentum and energy equations yields an additional set of stresses, the *Reynolds stresses*, in the momentum equation as well as additional energy advection terms in the energy equation. In their final form, the time-averaged equations for turbulent flow and heat transfer are as follows:

$$\rho_f \left(\frac{\partial \bar{u}_i}{\partial t} + \bar{u}_j \frac{\partial \bar{u}_i}{\partial x_j} + \frac{\partial \overline{u_i' u_j'}}{\partial x_j} \right) = \rho_f g_i - \frac{\partial \bar{P}}{\partial x_i} + \mu_f \frac{\partial^2 \bar{u}_i}{\partial x_j \partial x_j} \tag{2.6}$$

and

$$\rho_f c_f \left(\frac{\partial \bar{T}}{\partial t} + u_j \frac{\partial \bar{T}}{\partial x_j} + \frac{\partial \overline{T' u_j'}}{\partial x_j} \right) = k_f \frac{\partial^2 \bar{T}}{\partial x_j \partial x_j}. \tag{2.7}$$

The first of the last two equations is often called the *Reynolds-averaged Navier–Stokes (RANS)* equation. It is used in the so-called *RANS models* for turbulence. The time-averaged terms in the last two equations are modeled by closure equations, which typically emanate from analysis, supplemented by experimental or other numerical data. Two simple closure equations for the two terms emanate from Prandtl's *mixing length* theory, which is often called the *zero-equation model* and has been very successful in modeling channel flows. The final expressions from the mixing length theory may be written as follows in the x, y, z system of coordinates (z is along the axis of symmetry of the channel flow):

$$\overline{u'_x u'_y} = \ell^2 \left|\frac{\partial \bar{u}_x}{\partial y}\right| \frac{\partial \bar{u}_x}{\partial y} = 0.41 y^2 \left|\frac{\partial \bar{u}_x}{\partial y}\right| \frac{\partial \bar{u}_x}{\partial y} \tag{2.8}$$

and

$$\overline{u'_x T'} = \ell^2 \left|\frac{\partial \bar{T}}{\partial y}\right| \frac{\partial \bar{u}_x}{\partial y} = 0.41 y^2 \left|\frac{\partial \bar{T}}{\partial y}\right| \frac{\partial \bar{u}_x}{\partial y}. \tag{2.9}$$

The last two expressions have been applied to turbulent boundary layer flows, such as the ones formed over flat plates and inside pipes and channels, and have produced accurate results. However, the equations have proven to be rather inaccurate when used with more complex, three-dimensional flows. For this reason, other turbulent models have been formulated, such as the *k–ε model* and three-, six-, or nine-equation models (Warsi 1993). Of these, the *k–ε* model and its several spinoffs are frequently used in most of the commercially available codes. These models use the concept of *eddy viscosity*, which is defined as

$$\mu_T = C \rho_f \frac{k^2}{\varepsilon}, \tag{2.10}$$

where C is a constant, with typical value $C = 0.09$; k is the kinetic energy of the velocity fluctuations, $k = 0.5 \Sigma(u'_j{}^2)$; and ε is the rate of dissipation of the turbulent fluctuations. The *k–ε* model and similar models include two additional differential equations for the variables k and ε that need to be solved in conjunction with the Navier–Stokes equations (Warsi 1993).

The *large eddy simulation (LES)* method is often used in order to resolve for the larger vortices/eddies in the turbulence flow field. The LES method averages the smaller turbulent structures, especially those close to the boundaries. The LES method uses a decomposition of the velocity field that allows the larger eddies to be computed:

$$u_i(t) = \bar{u}_i + \tilde{u}_i(t) + u'_i(t), \tag{2.11}$$

with the second term representing the resolved large eddy structures, which are numerically computed. An averaging process similar to the *k–ε* model or simpler

Fig. 2.3 Turbulent energy
spectrum and computational
capability of the RANS, LES,
and DNS models. The *dashed
arrows* indicate the ranges of
the resolved lengthscales

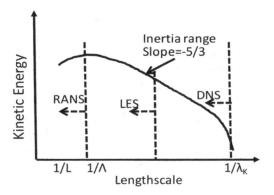

algebraic models (Smagorinsky 1963) is used for the averaged stresses that emanate
from the third term. The LES model gives a great deal of spatial and temporal
information about the large flow structures that appear in engineering systems
and has become a very promising and accurate computational technique in
multiphase flows.

The *direct numerical simulation* (*DNS*) method solves the complete time-
dependent Navier–Stokes equations and does not require a turbulence model. The
DNS method, typically, resolves all eddy sizes and uses a very fine grid, which
implies a high amount of computational resources. For most engineering
computations at moderate or high Reynolds numbers, the fine grid requirement
makes DNS computations prohibitive for large systems. Oftentimes, the detailed
information provided by the DNS is also unnecessary for engineering design and
optimization purposes. The main advantage of the DNS computations is that they
are very accurate; they do not depend on assumptions or other closure equations and
may be used to develop closure equations for the LES and RANS models.

Figure 2.3 is a diagram of a developed turbulent kinetic energy spectrum and the
lengthscales resolved by each method for turbulence modeling. In this figure, L is
the characteristic lengthscale of the engineering system, Λ is the turbulence integral
lengthscale, and λ_K is the Kolmogorov microscale.

It must be noted that *hybrid* or combination methods, such as the RANS–LES
method, have also been used in numerical applications. These methods are a
compromise between the information gained by the solution of the model and the
computational resources devoted to the solution.

For the determination of the heat transfer processes in most models, first, the
velocity field is determined, and, secondly, the temperature field is computed using
Eq. (2.7). The local heat transfer is then computed using the fundamental conduction equation, and the space-averaged heat transfer is computed by spatially
integrating the local heat transfer.

2.3.3   Transitional Flows

Transitional flows are also characterized by velocity fluctuations, which start as two-dimensional and develop to become three dimensional. The turbulent flow structure is not well developed in transitional flows, and there is no apparent relationship and delineation between small and large eddies. Transitional flow is a difficult regime to simulate, primarily because a great deal is unknown in this regime. Both Eulerian models and Eulerian–Lagrangian models for the large vortices have been used in transitional flow simulations. In most of the engineering applications, transitional flows are modeled the same way as turbulent flows.

2.4   Modeling of Particulate Systems

Unlike other multiphase flow systems, where the phases flow in different and very complex regimes (e.g., bubbly, churn, or plug flow in gas–liquid systems), particulate flow systems are composed of dispersed particles or clusters of particles. For modeling purposes, the carrier fluid is always modeled as a continuum in an Eulerian way. The particulate phase or phases are modeled either in a Lagrangian or a Eulerian framework. An implicit condition that must be satisfied for the dispersed phase to be treated as a continuum in an Eulerian way is that the average interparticle distance must be significantly less than the size of the computational grid ($r_{ij} \ll \Delta x$). This imposes a lower limit on the flow features that may be resolved by the computational method and also implies that a large number of particles must be present in every cell of the grid. Figure 2.4 depicts two computational cells (a and b) where this implicit assumption is satisfied, one cell (c) where the continuum assumption is clearly not satisfied, and a fourth cell (d) which is a borderline case. If the Eulerian model were to be used for the description of the flow in cell (d), the computational results must be well validated.

Fig. 2.4 Cells (a) and (b) have a sufficient number of particles for the particulate phase to be modeled as a continuum. Cell (c) may not be modeled as a continuum. Cell (d) is a borderline case: If the particulate phase in cell (d) were modeled as a continuum, in a Eulerian way, the results of the model would have to be well validated

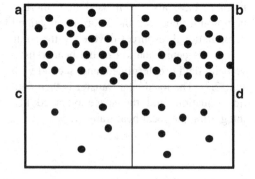

2.4.1 Eulerian Homogeneous Model

A rather simple method to model a particulate system is to use the assumption that the fluid–particles mixture is homogeneous. The thermodynamic properties of the *homogeneous mixture* are given in terms of the volumetric fraction, ϕ, of the dispersed phase as follows (see Sect. 4.4 for more details):

$$\rho_{\mathrm{m}} = (1 - \phi)\rho_{\mathrm{f}} + \phi\rho_{\mathrm{s}} \quad \text{and} \quad c_{\mathrm{m}} = \frac{1}{\rho_{\mathrm{m}}}[(1 - \phi)c_{\mathrm{f}}\rho_{\mathrm{f}} + \phi c_{\mathrm{s}}\rho_{\mathrm{s}}]. \tag{2.12}$$

The momentum and energy equations for the homogeneous mixture become

$$\frac{\partial \rho_{\mathrm{m}} u_{\mathrm{m}i}}{\partial t} + \frac{\partial \rho_{\mathrm{m}} u_{\mathrm{m}j} u_{\mathrm{m}i}}{\partial x_j} = \rho_{\mathrm{m}} g_i - \frac{\partial P}{\partial x_i} + \mu_{\mathrm{m}} \frac{\partial^2 u_{\mathrm{m}i}}{\partial x_j \partial x_j} \tag{2.13}$$

and

$$\frac{\partial \rho_{\mathrm{m}} c_{\mathrm{m}} T_{\mathrm{m}}}{\partial t} + \frac{\partial \rho_{\mathrm{m}} c_{\mathrm{m}} u_{\mathrm{m}j} T_{\mathrm{m}}}{\partial x_j} = k_{\mathrm{m}} \frac{\partial^2 T_{\mathrm{m}}}{\partial x_j \partial x_j}. \tag{2.14}$$

It must be noted that the velocity u_{m} and temperature T_{m} are the space-averaged variables of the particulate mixture and that the two may be different than the corresponding variables of the carrier fluid and of the particulate phase. Also it must be noted that, while the thermodynamic properties of a mixture are well-defined as shown in Eq. (2.12), there is no agreement as to the definition of the transport properties of the mixture, μ_{m} and k_{m}. As will be further elaborated in Sect. 4.5, there is strong experimental evidence that the transport properties of the mixture depend on the distribution of the particles in the mixture. As a result, there is not a rigorous and accurate definition of μ_{m} and k_{m}. Modelers use ad hoc assumptions for the transport properties and, oftentimes, assume that they are equal to the corresponding properties of the carrier phase, μ_{f} and k_{f}. The epistemic uncertainty in the definition of the transport properties adds to the overall uncertainty of the computed results.

The homogeneous model is based on the average properties of the two phases and provides information only on these averages. It does not distinguish between particles and carrier fluid and does not answer questions, such as what is the relative velocity of the particles or what is the effect of the particulate phase on the overall heat transfer characteristics of the mixture. Because separate and more detailed information for the behavior of the two phases is desired, another Eulerian model, the two-fluid model, is often used for the modeling of particulate mixtures.

2.4.2   Eulerian, Two-Fluid Model

The *two-fluid* model, which is also called the *Eulerian point-source* model, treats the carrier fluid and the particles as two distinct continua that occupy the same volume. The two continua are governed by their own conservation equations (mass, momentum, and energy). The interactions of the two continua are modeled by source terms, which are added to the governing equations. For example, the momentum equation of the carrier fluid contains an additional force term that represents the drag exerted by the particles on this fluid. Similarly, the mass and energy conservation equations of the carrier fluid include terms that represent the sublimation/evaporation of particles and the energy transfer from the particles to the fluid, respectively. These terms need to be modeled, and typically, empirical equations are used for their modeling. The complete system of equations for the point-source model, applied to Newtonian fluids, is as follows:

A. Mass conservation for the fluid and the particulate phase:

$$\frac{\partial[(1 - \phi)\rho_f]}{\partial t} + \frac{\partial[(1 - \phi)\rho_f u_j]}{\partial x_j} = J. \tag{2.15}$$

$$\frac{\partial(\phi\rho_s)}{\partial t} + \frac{\partial(\phi\rho_s v_j)}{\partial x_j} = -J. \tag{2.16}$$

B. Momentum conservation for the fluid and the particulate phase:

$$\frac{\partial[(1 - \phi)\rho_f u_i]}{\partial t} + \frac{\partial[(1 - \phi)\rho_f u_j u_i]}{\partial x_j} = (1 - \phi)\left[\rho_f g_i - \frac{\partial P}{\partial x_i} + \mu_f \frac{\partial^2 u_i}{\partial x_j \partial x_j}\right] + F_i + J v_i. \tag{2.17}$$

$$\frac{\partial(\phi\rho_s v_i)}{\partial t} + \frac{\partial(\phi\rho_s v_j v_i)}{\partial x_j} = \phi\left[\rho_s g_i - \frac{\partial(P + P_c)}{\partial x_i} + \mu_s \frac{\partial^2 v_i}{\partial x_j \partial x_j}\right] - F_i - J v_i. \tag{2.18}$$

C. Energy equation for the fluid and the carrier phase:

$$\frac{\partial[(1 - \phi)\rho_f c_f T_f]}{\partial t} + \frac{\partial[(1 - \phi)\rho_f c_f u_j T_f]}{\partial x_j} = (1 - \phi)k_f \frac{\partial^2 T_f}{\partial x_j \partial x_j} + q. \tag{2.19}$$

$$\frac{\partial(\phi\rho_s c_s T_s)}{\partial t} + \frac{\partial(\phi\rho_s c_s v_j T_s)}{\partial x_j} = \phi k_s \frac{\partial^2 T_s}{\partial x_j \partial x_j} - q. \tag{2.20}$$

The mass source term, J, represents the mass transferred to the carrier fluid from the particles as a result of evaporation, sublimation, or chemical reactions.

The same term multiplied by the particle velocity appears in the momentum equations to represent the momentum transferred as a result of this mass exchange between the phases. In addition, the force term, F_i, also appears in the momentum equations to represent the hydrodynamic force between fluid and particles, such as drag and lift. Finally, the heat source term, q, represents the entire enthalpy transfer per unit volume from the particles to the carrier fluid and includes the latent heat of evaporation or sublimation, h_{fg}. The pressure term, P_c, represents the particle collisions and may be neglected if collisions are unimportant. It must be noted that terms, which are typically of significantly lesser orders of magnitude, such as the viscous dissipation term, have been omitted from the energy equation of the carrier fluid. As with the homogeneous model, the transport coefficients of the particulate phase, k_s and μ_s, have been assumed to be constant and need to be defined from empirical expressions. Oftentimes, these terms are assumed to be equal to the corresponding transport coefficients of the carrier fluid.

The equations in the point-source or two-fluid model are an extension of the governing equations of Newtonian fluids. As such, the system of equations of the model is robust and may be solved numerically by several of the algorithms that have been developed for the single-phase CFD. The accuracy of the model depends very much on the accuracy of the closure equations that are used for the interaction terms and the transport coefficients. For this reason, a great deal of computational and experimental work is being done to refine the closure equations.

Another source of uncertainty for the two-fluid models is the specification of the particles–wall boundary conditions that apply to the PDEs of the particulate phase: While the no-slip condition is routinely applied to solid walls as an accurate and time-tested boundary condition for fluids, it is not intuitive that the boundary condition for the solid phase should also be the no-slip condition. Actually, several experimental and DNS numerical studies have proven that there is significant slip of the particulate phase at the solid boundaries (Davis et al. 2011). The wall slip has to be given by a closure equation that is produced from experimental data or detailed computations. Two complications related to the specification of the particles–wall boundary condition are:

(a) The boundary condition for the particulate phase is defined at the plane where the centers of the particles are located when in contact with the wall. For spherical particles, this distance is one radius from the wall. Again, if particles of several sizes are present, or if the shapes of the particles are nonspherical, it is not clear where exactly the boundary condition for the solid phase should be applied. One way to address the second difficulty is to define several particulate phases, one phase for each size of particles (Mostafa and Elghobashi 1985). In the case of the continuous distribution of particle sizes, the phase may be defined for a range of sizes. While this practice may simplify the application of the boundary conditions, it increases significantly the number of PDEs that are to be solved, and if many phases need to be defined and their variables computed, the method becomes impractical.

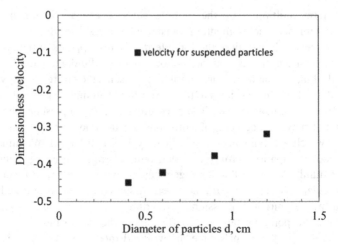

Fig. 2.5 Dimensionless vertical velocity of spherical particles at the plane where the boundary condition is defined

(b) The particle wall slip depends on the size of the particles. When the particulate phase is composed of particles with different sizes and shapes, it is neither known nor is it intuitive at all, what is the form of the velocity slip function at the solid boundary. As may be seen in Fig. 2.5, which depicts the dimensionless vertical velocity of particles at a plane one radius away from a vertical solid boundary, the particle slip at the vertical wall is finite and depends on the size of the particles. This difficulty may be addressed by conducting appropriate experimental or computational (DNS) studies near walls that will yield meaningful and reliable boundary conditions for the particulate phase.

2.4.3   Lagrangian, Point-Source Model

When dealing with discrete particles, it is more intuitive and physically meaningful to use a Lagrangian description with the center of coordinates at the center of gravity of the particles. The Lagrangian point-source models treat the particles as points that move in the flow field and are sources of mass, momentum, and energy for the fluid. The carrier fluid is treated as a continuum in an Eulerian way, and the velocity and temperature fields for the fluid are obtained from the solution of the PDEs in Eqs. (2.15), (2.17), and (2.19). The particle trajectories and temperatures are obtained by the solution of the ODEs, which emanate from the equation of motion and energy for particles:

$$m_s \frac{dv_i}{dt} = \sum (F_{Bi} + F_{Si} + F_{Ci}) \tag{2.21}$$

and

$$m_s c_s \frac{dT_s}{dt} = \sum \dot{Q}. \tag{2.22}$$

The forces in Eq. (2.21) include the body force, F_B, and surface or hydrodynamic force, F_S, on the particles as well as any interaction or collision forces with other particles and with the boundaries, F_C. Among the components of the hydrodynamic force are the steady drag, the added mass, the history, and the lift force, which are given in more detail in Sect. 1.3.6. Similarly, for the heat transfer, $\dot{Q}$, the steady convection as well as the history term must be included as described in Sect. 1.4.4. When the particles are considered as points, the angular momentum equation becomes meaningless. However, within the analytical framework of this model, particles of finite size may be considered. In this case, the angular momentum equation for the particles becomes:

$$I_p(\rho_s - \rho_f) \frac{d\omega_k}{dt} = -\rho_f \int_S e_{ijk}(x_j - x_{cj})(F_{Sk} + F_{Ck})dS, \tag{2.23}$$

where the integral is computed around the surface of the particle and includes the friction forces; the tensor e_{ijk} defines the vector product (cross product) in the system of coordinates, i,j,k; and the point x_c represents the center of the particle. If the surface forces on the particle are given in terms of closure equations, this integral is equal to the sum of all the cross products of forces and the positions of the points of application.

The solution of the set of Eqs. (2.21), (2.22), and (2.23) is accomplished on a Lagrangian system of coordinates that follows the centers of the particles, usually by a time-marching method. For large numbers of particles, computational "parcels" are often used. Each parcel represents a number of particles with the same characteristics, such as shape, size, and density. A computational restriction in this case is that the size of the entire parcel should be smaller than the size of the computational grid.

For intermediate and dense flows, particle collisions play an important role in the determination of the trajectories of particles. Simple, deterministic collision models that emanate from first principles (momentum conservation and partial mechanical energy dissipation during the collision) are not sufficiently accurate to describe the particle interactions, especially where nonspherical particles and particles of different sizes are present. For this reason, probabilistic collision models have been proposed and are being used. The collision models are examined in more detail in Sect. 2.4.6.

The Lagrangian, point-source model is robust and relatively easy to implement, especially when the particle motion and energy exchange does not significantly influence the velocity and temperature fields of the fluid. A special case of the application of the model is the so-called *Monte Carlo (MC) simulations*, which were originally developed to simulate the effects of fluid turbulence and

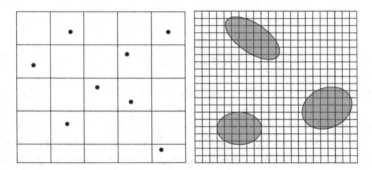

Fig. 2.6 The point-source model may handle a large number of particles, which must be smaller than the size of the grid. The resolved-particle model may only handle a small number of particles, and its grid must be significantly smaller than the size of the particles

time-varying temperature fields on the motion of particles (Gosman and Ioannides 1983) and on the heat transfer of particles (Michaelides et al. 1992). With MC simulations, the average carrier fluid velocity and temperature fields are first solved assuming there are no particles. Secondly, a probabilistic approach to model the velocity and temperature fluctuations of the fluid is used, typically from a known random distribution. A large number of particles are inserted in the flow field, and their momentum and heat exchange with the carrier fluid are computed using Eqs. (2.21), (2.22), and (2.23). The ensemble average of all the particles is computed. According to the ergodic hypothesis, the ensemble-averaged properties are equal to the time-averaged properties of the particles.

The drawback of the simple MC simulations is that the effects of the particles on the fluid velocity and temperature fields, as well as the particle–particle interactions, are inherently neglected. The way the method is commonly applied renders the model a one-way interaction model. Because of this, MC simulations may be applied only to dilute flows and only give a qualitative representation of the average behavior of particles in denser flows or in flows where some regions have higher particle concentrations.

2.4.4 Lagrangian, Resolved-Particle Model

This is the type of model used in direct numerical simulations (DNS). In the resolved-particle model, the details of the solution of the carrier phase velocity and temperature fields are such that it is feasible to determine the particle–fluid interactions from first principles. In this model the carrier phase numerical grid size is significantly smaller than the size of the particles ($\Delta x \ll \alpha$). Figure 2.6 contrasts the carrier phase grid size of this model with that of the point-source model. It is apparent that the resolved-particle model requires significantly higher computational resources for comparable numbers of particles. The solution of the carrier phase governing equations determines the pressure, velocity, and temperature fields around each particle. The mass flux, the hydrodynamic surface force, and the rate of heat transfer between each particle and the fluid may be determined by integrating the concentration, stress, and temperature gradient fields over the surface of the particles, A_p, as follows:

$$J = \int_{A_s} \rho_v D_{vf} \frac{\partial c_v}{\partial x_j} n_j dA_s \tag{2.24}$$

where D_{vf} is the diffusion coefficient of the vapor stemming from the particles in the carrier fluid because of sublimation or evaporation; ρ_v is the vapor density; c_v is the volumetric concentration of the vapor in the carrier fluid; and n_j represents the outward normal vector to the particle surface, A_s. The hydrodynamic force is calculated from the expression:

$$F_{s,i} = \int_{A_s} \left[-P\delta_{ij} + \mu_f \left(\frac{\partial u_i}{\partial x_j} + \frac{\partial u_j}{\partial x_i} \right) \right] n_j dA_s \tag{2.25}$$

where δ_{ij} is the Kronecker delta. Finally, the rate of heat that enters[1] the particle is

$$\dot{Q} = \int_{A_s} \left(-k_f \frac{\partial T_f}{\partial x_j} \right) n_j dA_s. \tag{2.26}$$

The equation for the rotational motion of the particle in this model is the same as Eq. (2.23). Under this model there is no need for closure equations to account for the particle–fluid interactions. Of course, the resolution of the fields around the particles and the condition $\Delta x \ll \alpha$ implies that a very fine grid must be used in the resolved computations. A great deal of computational resources must be used even when the motion and energy exchange from a moderate number of particles is considered. Because of this, the resolved method is not suitable when a large number of particles need to be simulated in the system. This precludes the application of the DNS model to large engineering systems, such as fluidized bed reactors, which contain a very large number (of the order of 10^{10}) of discrete particles. Actually, such detailed information on the behavior of individual particles is not necessary for the design of large engineering systems.

The main advantage of the resolved-particle or DNS model is that it does not require empirical closure equations for the fluid–particle interactions. These are determined from first principles. The only empirical information required by the resolved-particle model is related to the collision of the particles. With a suitable collision scheme, this model determines accurately the behavior of all particles. When the grid of this model is very fine, in order to provide high accuracy results

---

[1] In Eqs. (2.22) and (2.26) we follow the thermodynamic convention: Heat that enters the system (particle) is positive.

for the particle–fluid interactions, then the DNS model itself may be used for the development of other, needed closure equations of these interactions. This includes the development of closure equations for the drag and convective heat transfer coefficients, C_D and h_c, from certain irregularly shaped particles, for which closure equations are not currently available. Such closure equations for the fluid–particle interactions may be further used to improve the accuracy of the Eulerian and the point-source models, thus enabling the simulations of large number of particles and realistic engineering systems. Therefore, the detailed information obtained from the resolved-particle model may be used to provide accurate information that feeds into models for large-scale engineering systems, where global information is required and the behavior of individual, separate particles is not of interest.

2.4.5   The Probability Distribution Function Model

The probability distribution function (PDF) method has been developed to handle simultaneously the flow turbulence and the behavior of small particles in a turbulent flow field. The origins of this method are in the kinetic theory of gases and can be traced to the studies by Maxwell and Boltzmann. Probabilistic methods were developed for the dense flow or granular materials, where the behavior of the particulate system is dominated by the collisions (Jenkins and Richman 1985; Ding and Gidaspaw 1990). These models apply to dense particulate systems, where the volumetric fraction is higher than 10% and the influence of the interstitial gas on the particle transport properties is almost negligible. Seeking an extension to lower volumetric fractions and dilute particulate systems, several researchers developed similar approaches that are based on the PDF equations for particles with additional closure equations, which account for the mass transfer, the hydrodynamic force, the heat transfer, fluid turbulence, and interparticle collisions.

An integral part of the kinetic theory models is the existence of a general equation that contains, implicitly or explicitly, terms, which yield the continuum description of the underlying medium. In the case of the kinetic theory of gases, the general equation is the Maxwell–Boltzmann equation for the probability distribution of the molecular velocities. In the case of particulate mixtures, the general equation is the *PDF equation*. The characteristic of the general PDF equation is that it may yield by a formal mathematical way the continuum equations for the flow and heat transfer of the carrier gas and the particles and, also, the natural boundary conditions that are observed near the walls—the near wall behavior of particles. Several PDF equations and models have been developed in the 1990s and have been used successfully to derive continuum conservation equations for particulate turbulent flows and heat transfer. Morioka and Nakajima (1987), Reeks (1991), and Zaichik and Vinberg (1991) were among the first to propose PDF equations for the treatment of the statistical averages and the behavior of multiphase systems.

Let us consider the motion of a dilute particulate system, where the particles exchange mass momentum and energy with the carrier fluid. We will denote by $X(t)$

the phase-space vector of a single particle as it moves through the phase space. The phase space in this case has 8 dimensions—three for the position, three for the velocity, one for the instantaneous mass, and one for the instantaneous temperature—and may be written at the instant of time t formally as

$$X(t) = [\vec{v}, \vec{x}, m_s, T_s]. \tag{2.27}$$

The time derivative of the phase-space vector may be obtained explicitly:

$$\dot{X}(t) = \left[\dot{\vec{v}}, \dot{\vec{x}}, \dot{m}_s, \dot{T}_s\right] = \left[\dot{\vec{v}}, \vec{v}, \dot{m}_s, \dot{T}_s\right]. \tag{2.28}$$

The last equation implies that the phase-space vector contains implicitly or explicitly information on the equation of motion of the particles, the heat transfer equation, and the mass transfer equation of the particles in the carrier fluid.

In analogy with the kinetic theory of gases, the number of particles in an elemental volume of the phase space $d^n X$ located at X will be given by the product of the phase-space density function, $W(X,t)$, and the elemental volume $d^n X$. The fundamental number conservation equation may be applied to the phase space, to yield the condition:

$$\frac{\partial W}{\partial t} + \frac{\partial}{\partial X}\left[W\dot{X}\right] = 0. \tag{2.29}$$

Because turbulence is developed in the carrier fluid, the time derivative of the phase-space vector, $\dot{X}(t)$, has a time-dependent component, which may be assumed to be random. Therefore, one may assume a number of realizations of the phase-space vector, $X(t)$, at a given instant of time t. One may take the *ensemble average* of all the realizations of the phase-space density function, W, which will be denoted as $<W>$. The equation for $<W>$ is the PDF equation of this problem and may be obtained by ensemble averaging the last conservation equation. In the case of the dilute system of particles considered here, one may decompose the result to write explicitly its PDF equation as follows:

$$\frac{\partial \langle W \rangle}{\partial t} + \left(\frac{\partial \langle \dot{m}_s \rangle}{\partial m_s} + \frac{\partial \langle \dot{T}_s \rangle}{\partial T_s} + \frac{\partial \vec{v}}{\partial \vec{x}} + \frac{\partial \langle \dot{\vec{v}} \rangle}{\partial \vec{v}}\right) \langle W \rangle + \frac{\partial \langle \dot{m}'_s W \rangle}{\partial m_s} + \frac{\partial \langle \dot{T}'_s W \rangle}{\partial T_s} + \frac{\partial \langle \dot{\vec{v}}' W \rangle}{\partial \vec{v}} = 0, \tag{2.30}$$

where it was recognized that $\vec{v} = \dot{\vec{x}}$.

It is apparent that the PDF equation must be supplemented with expressions for $\dot{m}_s$, $\dot{T}_s$, $\dot{\vec{v}}$, etc. Such expressions are generated from the equation of motion and the energy equation of single particles. For example, the expressions for the particle acceleration vector and the rate of temperature change for dilute particulate flows may be obtained directly from the appropriate expressions in Sects. 1.3 and 1.4. For more details of this method of modeling and some of the results that may be obtained, Reeks and Simonin (2006) and Simonin (2001) provide excellent reviews on the subject.

2.4.6 Particle Collisions

Particle collisions are infrequent in dilute flows and for this reason are neglected. Collisions increasingly influence the motion and heat transfer of particulates with increasing concentration. Except in very dense mixtures with $\phi > 45\%$, particle collisions are considered *binary*. The term implies that collisions between more than two particles are infrequent enough to be neglected. This stipulation simplifies considerably the analytical treatment of the collision processes and the effects of the collisions on the transport properties of the mixture. When multiparticle collisions become dominant, as in very dense particulate flows, the flows are characterized as *granular flows*. Usually fluid–particle interactions and inertia effects are neglected in the treatment of granular flows.

Rigid particles collide and separate. Liquid drops may collide and either separate with no other change, or break up, or coalesce. In the last two cases, the collision process is dominated by the surface deformations and surface force effects. Inter-particle collisions occur during a finite amount of time, which is, in general, much shorter than the characteristic times of the particles, τ_M and τ_{th}. During the short collision process, interaction forces are developed between the particles, which are by far greater than the hydrodynamic forces between the particles and the fluid. Depending on the surface properties and the type of collision, the particles may also slide at the contact surface. A sliding friction force is thus developed, which is normally modeled using Coulomb's friction law. Deformations of the surfaces of the particles occur during the collision process. In most cases, the deformations are assumed to be negligible in comparison to the interparticle distance. Therefore, the interparticle distance remains constant during the collision process, and the contact may be assumed to occur at a single point, where the interparticle force is applied.

Two mathematical models to describe the collisions of particles have been developed: the *hard-sphere model* and the *soft-sphere model*. In the hard-sphere model, the impulses of all the forces between the colliding particles are assumed to be constant and are given in an integral form. This model lumps all the effects of the collision process into a single variable: the impulse produced by the interparticle force during the entire collision process. In the soft-sphere model, the governing equations are given in differential form. The magnitude of forces and moments vary during the collision process. Newton's second law determines the particles' velocity changes due to these transient forces and moments.

Figure 2.7 shows schematically the hard-sphere collision process. The initial velocities of the particles are denoted by the subscript 0, and the impulse force, which is developed during the collision, is denoted as $\vec{F}$. Elementary mechanics

Fig. 2.7 Instantaneous forces between two colliding solid spheres

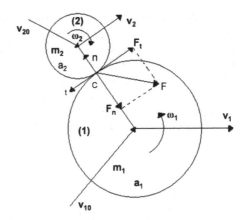

theory proves that the rectilinear and angular velocities of the two particles after the collision process are given by the following expressions:

$$m_1(\vec{v}_1 - \vec{v}_{10}) = m_2(\vec{v}_2 - \vec{v}_{20}) = -\vec{F} = \int_0^{\delta t} \vec{f} dt \tag{2.31}$$

$$I_1(\vec{\omega}_1 - \vec{\omega}_{10}) = \alpha_1 \vec{n} \times \vec{F}, \quad I_2(\vec{\omega}_2 - \vec{\omega}_{20}) = \alpha_2 \vec{n} \times \vec{F}.$$

It must be noted that $\vec{F}$ denotes the impulse of the all the forces acting on the particles during the entire duration of the collision time, δt. For spherical particles, the moment of inertia I is equal to $0.4 \, ma^2$. Since the impulse of the force, $\vec{F}$, cannot be determined from the first principles of mechanics, the hard-sphere model for collisions makes use of the relative motion of the two particles and their material properties to derive expressions for the particle velocities at the end of the collision process. Let us decompose the impulse, $\vec{F}$, into two components: the first along the line of the centers of the two particles, which is normal to the surfaces at the point of contact, and the second in the perpendicular direction of the line between the centers, which is the tangential direction at the point of contact. Hence,

$$\vec{F} = F_n \vec{n} + F_t \vec{t}. \tag{2.32}$$

The normal relative velocities of the two particles, before and after the collision process, are related by a restitution coefficient, e_r:

$$\vec{n} \cdot (\vec{v}_1 - \vec{v}_2) = -e_r \vec{n} \cdot (\vec{v}_{10} - \vec{v}_{20}) \quad \text{or} \quad \vec{n} \cdot \vec{w} = -e_r \vec{n} \cdot \vec{w}_0. \tag{2.33}$$

From the last three equations, one obtains the following expression for the normal component of the impulse force:

$$F_n = \frac{-m_1 m_2}{m_1 + m_2} (1 + e_r)(\vec{n} \cdot \vec{w}_0). \tag{2.34}$$

For the collision to occur, the normal component of the relative velocity of the particles must be in the direction of the vector $\vec{n}$. Hence, the last scalar (dot) product is positive. Since the restitution coefficient is also positive, the last equation implies that $F_n < 0$. Hence, the normal force is directed inward, that is, in the direction defined from the point of the collision to the center of the particle.

If the particles slide during the collision process and the coefficient of friction is denoted by f_f, then Coulomb's law of friction is applied ($F_t = f_f F_n$) to yield the tangential component. The condition for sliding to occur is (Crowe et al. 1998)

$$F_t > -\frac{2}{7}\frac{m_1 m_2}{m_1 + m_2}|\vec{w}_{0tc}| \quad \text{or} \quad \frac{\vec{n} \cdot \vec{w}_0}{w_{0tc}} < \frac{2}{7 f_f(1 + e_r)}, \tag{2.35}$$

where the vector $\vec{w}_{0tc}$ is the initial tangential velocity at the point of contact.

Hence, the linear and angular velocities of the two particles after the sliding collision process are given by the following expressions:

$$\vec{v}_1 = \vec{v}_{10} - (\vec{n} \cdot \vec{w}_0)(1 + e_r)\frac{m_2}{m_1 + m_2}(\vec{n} - f_f \vec{t})$$

$$\vec{v}_2 = \vec{v}_{20} + (\vec{n} \cdot \vec{w}_0)(1 + e_r)\frac{m_1}{m_1 + m_2}(\vec{n} - f_f \vec{t})$$

$$\vec{\omega}_1 = \vec{\omega}_{10} + \frac{5}{2\alpha_1}(\vec{n} \cdot \vec{w}_0)f_f(1 + e_r)\frac{m_2}{m_1 + m_2}(\vec{n} \times \vec{t}) \tag{2.36}$$

$$\vec{\omega}_2 = \vec{\omega}_{20} + \frac{5}{2\alpha_2}(\vec{n} \cdot \vec{w}_0)f_f(1 + e_r)\frac{m_1}{m_1 + m_2}(\vec{n} \times \vec{t}).$$

If the sliding motion stops during the hard-sphere collision process, the condition of Eq. (2.35) is not satisfied. In this case, at the end of the collision process, the relative velocity of the particles is zero, and the expressions for the particle velocities after the collision are

$$\vec{v}_1 = \vec{v}_{10} - \frac{m_2}{m_1 + m_2}\left[(1 + e_r)(\vec{n} \cdot \vec{w}_0)\vec{n} + \frac{2}{7}|\vec{w}_{0tc}|\vec{t}\right]$$

$$\vec{v}_2 = \vec{v}_{20} + \frac{m_1}{m_1 + m_2}\left[(1 + e_r)(\vec{n} \cdot \vec{w}_0)\vec{n} + \frac{2}{7}|\vec{w}_{0tc}|\vec{t}\right]$$

$$\vec{\omega}_1 = \vec{\omega}_{10} - \frac{5}{7\alpha_1}|\vec{w}_{0tc}|\frac{m_2}{m_1 + m_2}(\vec{n} \times \vec{t}) \tag{2.37}$$

$$\vec{\omega}_2 = \vec{\omega}_{20} - \frac{5}{7\alpha_1}|\vec{w}_{0tc}|\frac{m_1}{m_1 + m_2}(\vec{n} \times \vec{t}).$$

The expressions for the velocities of the two particles at the end of the collision process may be used as closure equations to determine the effect of the collisions on the dynamics of a particulate mixture.

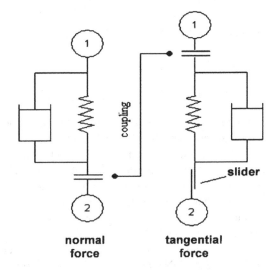

Figure 2.8 depicts the schematic diagram of the soft-sphere collision model. The
basic premise of the soft-sphere model is that the interparticle force is variable
during the collision process. This model calculates the instantaneous value of the
collision force, using Newton's second law. The soft-sphere model presumes that
the colliding particles overlap by a small distance, δ, which is very small in
comparison to the particles' dimensions. The overlapping distance is decomposed
into a normal, δ_n, and a tangential component, δ_t. These two components are
calculated from the initial strength of the impact between the particles and the
stiffness of the particles. The force model includes simple elements from solid body
dynamics, such as springs, dash pots, friction sliders, rollers, and latches. The
coupling between the normal and tangential components of the force is also
depicted in the figure. The stiffness coefficient, k_s; the damping factor, η_d; and the
friction factor, f_f, which may be calculated from the material properties of the
particles, are inputs to this model and are used to determine the normal and
tangential components of the instantaneous force. In the most general case, these
material properties have different values in the normal and tangential directions and
are expressed as functions of the Young's modulus, E_Y, and the Poisson ratio σ_P of
the materials. Thus, the components of the interparticle force for two spheres of
equal radii may be written by the following set of equations:

$$F_n = -\frac{\sqrt{2\alpha}E_Y}{3(1-\sigma_P)}\delta_n^{3/2} - \eta_{dn}\vec{w}\bullet\vec{n} \quad \text{and}$$

$$F_t = \frac{2\sqrt{2\alpha}E_Y}{2(1+\sigma_P)(2-\sigma_P)}\delta_n^{1/2}\delta_t - \eta_{dt}\left[\vec{w} - (\vec{w}\bullet\vec{n})\vec{n} + \alpha(\omega_i+\omega_j)\times\vec{n}\right]\bullet\vec{n} \quad (2.38)$$

$$\text{if } |F_t|<f_f|F_n| \quad \text{or} \quad F_t = -f_f|F_n| \quad \text{if} \quad |F_t|>f_f|F_n|.$$

Cundall and Strack (1979) recommended the following expressions for the damping coefficients:

$$\eta_{\mathrm{dn}} = \sqrt{m_{\mathrm{s}}\frac{\sqrt{2\alpha E_{\mathrm{Y}}}}{3(1-\sigma_{\mathrm{P}})}} \quad \text{and} \quad \eta_{\mathrm{dt}} = \sqrt{m_{\mathrm{s}}\frac{2\sqrt{2\alpha E_{\mathrm{Y}}}}{2(1+\sigma_{\mathrm{P}})(2-\sigma_{\mathrm{P}})}}\delta_{\mathrm{n}}^{1/2}. \qquad (2.39)$$

It must be noted that, according to the soft-sphere model, the interparticle force is instantaneous and that its numerical value varies during the collision process. The laws of mechanics are used in a differential form to determine the linear and angular velocity changes during the collision process. This is usually accomplished by a numerical method (Tsuji et al. 1993; Kartushinsky and Michaelides 2004). Also, both the hard- and soft-sphere models may be extended to multiparticle interactions, though this does not appear to be necessary for the modeling of discrete dispersed systems.

The collision models are developed independently of the larger computational models, e.g., a DNS or a two-fluid model, for the flow and heat transfer from particulate systems. When an interparticle or particle collision model becomes part of a larger computational model, it is important that the collision model does not disturb significantly the local characteristics of the larger numerical model. The disturbance might make the larger model numerically unstable. The sudden introduction of a large force locally, which accompanies the collision process, in both the hard- and soft-sphere models, may introduce computational instabilities in a larger numerical code. For this reason, in several numerical algorithms, the collision forces are introduced gradually and are often applied "gently" before the surfaces of the particles collide. In these algorithms, the collision models are modified so that a repulsive force starts acting on both particles when the interparticle distance or the distance of the particle from the wall is less than a predefined threshold distance, ζ, which is typically in the range $0.1\alpha < \zeta < 0.2\alpha$. Such a model has been proposed by Glowinski et al. (2001) and has been successfully used by several others including Feng and Michaelides (2004). According to this model, the repulsive force between two particles is given in terms of the distance between the centers of the two particles, $|x_i - x_j|$, by the following expression:

$$F_{ij}^{\mathrm{P}} = \begin{cases} 0, & |x_i - x_j| > \alpha_i + \alpha_j + \zeta \\ \frac{c_{ij}}{\varepsilon_{\mathrm{P}}}\left(\frac{|\bar{x}_i - x_j| - \alpha_i - \alpha_j - \zeta}{\zeta}\right)^2\left(\frac{x_i - x_j}{|x_i - x_j|}\right), & |x_i - x_j| \le R_i + R_j + \zeta \end{cases}. \qquad (2.40)$$

The parameter, c_{ij}, is the force scale, which for typical particulate systems is chosen to be equal to the buoyancy/gravity force on the particles. Of the other parameters, ε_{P} is the stiffness parameter for collisions, and α_i and α_j are the radii of the two particles. Glowinski et al. (2001) provided the justification and an extensive discussion on how to choose the stiffness and force parameters.

This collision technique allows particles to overlap even when the stiffness parameter c_{ij} is very large. The partial overlapping of particles will be significant when a large number of particles undergo a packing process, for example, in flow stagnation regions. The particles at the bottom, which have to bear the load of the particles above, will be subjected to the maximum overlapping, and this may distort the geometric characteristics of the computational domain. To counteract significant overlapping, one has to choose a higher value for the repulsive force when the collision scheme given by the above expression is used. To resolve this issue, Feng and Michaelides (2005) employed a new collision scheme that chooses the magnitude of the repulsive force by considering the following situations: Before the two particles contact, the repulsive force given by Eq. (2.40) is used; when the two particles start to overlap, a stronger spring force is applied. The latter is proportional to the overlapping distance of two particles and is significantly larger than the repulsive force with no overlapping. According to this approach, the collision force equation is modified to the following form:

$$
F_{ij}^{P} = \begin{cases} 0, & |x_i - x_j| > \alpha_i + \alpha_j + \zeta \\ \frac{c_{ij}}{\varepsilon_P} \left(\frac{|x_i - x_j| - \alpha_i - \alpha_j - \zeta}{\zeta} \right)^2 \left(\frac{x_i - x_j}{|x_i - x_j|} \right), & R_i + R_j < |x_i - x_j| \leq \alpha_i + \alpha_j + \zeta \\ \left(\frac{c_{ij}}{\varepsilon_P} \left(\frac{|x_i - x_j| - \alpha_i - \alpha_j - \zeta}{\zeta} \right)^2 + \frac{c_{ij}}{E_P} \frac{(\alpha_i + \alpha_j - |x_i - x_j|)}{\zeta} \right) \left(\frac{x_i - x_j}{|x_i - x_j|} \right), & |x_i - x_j| \leq \alpha_i + \alpha_j, \end{cases}
$$

$$(2.41)$$

where the parameter E_P is smaller than ε_P to ensure a much larger spring force, which will minimize the overlapping of the particles. The first term in the last equation is retained from Eq. (2.40) to ensure that the collision force will be continuous when the particles first touch. The advantage of this collision scheme is that it enables one to use a smaller repulsive force for particle sedimentation before the packing starts and a larger spring force that keeps the particles separated during the packing process, where a larger force is needed to keep the particles apart. This collision scheme may be used in dense as well as granular flows.

The modeling of particle collisions with a smooth wall may be accomplished in a manner similar to the interparticle collisions, by assuming that the wall is a very large particle. Either the hard- or the soft-sphere model may be used, and the conditions with $m_2 \gg m_1$ and $\alpha_2 \gg \alpha_1$ will yield the linear and angular velocities of the particle after the collision process. Alternatively, one may use the concept of the *image particle*. This is a second, fictitious particle, symmetrical with respect to the wall that moves with velocity, which is the image of the velocity of the particle (Glowinski et al 2001; Feng and Michaelides 2005). The fictitious collision between the two particles has the same effect as the collision with a solid wall.

A complication to the modeling process arises when the size of the particles is of the same order of magnitude as the wall roughness. Particles that approach a rough surface bounce in a direction that is determined by the local curvature and not by the

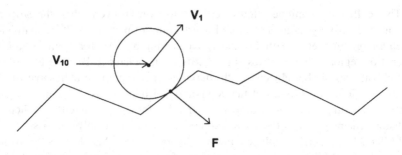

Fig. 2.9 Collision of a spherical particle with a rough surface—the reflection depends on the local geometry and not the overall shape of the surface

macroscopic shape of the surface. Hence, surface irregularities determine the direction of the bouncing particles as shown in Fig. 2.9. With collisions on a rough wall, it is neither possible nor desirable to simulate accurately the actual roughness of a wall surface. For this reason, statistical models for surfaces have been proposed that take into account the average features of the surface. These surface models use wavy patterns, random combinations of inclined planes, and random combination of pyramids and prisms (in three-dimensional simulations) arranged on a flat or rounded surface. Frank et al. (1993), Sakiz and Simonin (1999), Sommerfeld and Huber (1999), Sommerfeld (2003), and Taniere et al. (2004) have proposed such models for rough surfaces and used these models for particulate flow computations.

2.4.7 Droplet Collisions and Coalescence

Two viscous spheres, bubbles or drops, may coalesce when they are in close proximity or when they collide. The coalescence process is complex and depends on several variables including the size, surface tension, viscosity of the two phases, and the two velocity vectors (Manga and Stone 1993, 1995; Orme 1997). Most numerical methods do not handle coalescence from first principles and rely on closure equations and conditions for the entire process. Also, contrary to intuition, coalescence rarely occurs when two drops or bubbles interact (Qian and Law 1997). In most cases, when the paths of drops and bubbles intersect, the interaction causes collisions, which are similar to the collisions of solid particles, and separation at the end of the process. Experiments have shown that collisions between drops rarely occur in sprays, where the droplets move in almost parallel directions and that coalescence occurs only in dense regions, for example, near the orifice of an injector (Sirignano 1999).

Qian and Law (1997) conducted an extensive experimental study on the collision of two drops of equal size and showed that the behavior of the drops may be described in a plot of the Weber number, *We*, vs. the minimum dimensionless

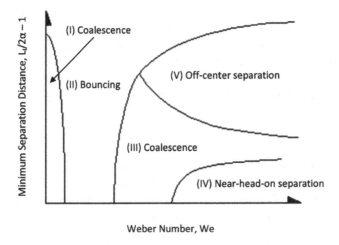

Fig. 2.10 Five coalescence regimes for the collision of two viscous spheres

separation distance, which is equal to $L_I/2\alpha - 1$. The outcomes of the collision process are two: coalescence or separation. The experimental results indicate that there are five distinct regimes for the collision process of the drops, three of which result in separation. The five regimes defined by Qian and Law (1997) are plotted qualitatively in Fig. 2.10. When two drops approach, the interstitial fluid between them stretches and becomes a thin film, with the pressure increasing locally. If the two drops approach slowly, the interstitial fluid film has time to drain, the surfaces of the drops touch, and coalescence occurs with minor deformation of the surfaces. This leads to collision regime I of Fig. 2.10, where the low values of *We* signify the low relative velocity of the drops. At higher initial relative velocity (higher *We*), the interstitial film does not have the time to drain, the higher pressure builds up quickly in the film, and the surfaces of the drops do not come in contact. The result is the deformation and repulsion/bouncing of the drops, which is depicted as regime II in Fig. 2.10. At even higher values of the relative velocity (and of *We*), the kinetic energy of the drops is high enough to forcibly expel the interstitial fluid film. The high kinetic energy also deforms substantially the two drops and finally causes their coalescence, as depicted in regime III. If the collisional kinetic energy of the drops is very high, the gaseous film is again drained, and the surfaces of the drops may touch temporarily. In this case, the system of the two drops has very high kinetic energy, which leads to vibrations and surface instabilities. The compound drop breaks up into two or more droplets. The experiments by Qian and Law (1997) distinguish two such regimes, denoted as IV and V in the figure, the first where the drops oscillate and undergo a reflective separation for a near head-on collision and the second where the drops stretch apart and undergo a stretching separation for off-center collisions.

Estrade et al. (1999) and Ashgriz and Poo (1990) provided a theoretical framework for the collision process of drops, which has resulted in analytical expressions for the description of the boundaries of regimes II and III, and of regimes III and IV.

Such analytical expressions, or the entire Fig. 2.10, may serve as closure equations in a numerical scheme for the determination of coalescence or separation. In the case of coalescence, the numerical computations continue by introducing a single drop in the computational domain and the colliding drops are taken off the computations. If the radii of the two colliding drops are α_1 and α_2, according to the volume conservation principle, the radius of the resulting drop after the coalescence, α_{12}, is

$$\alpha_{12} = \sqrt[3]{\alpha_1^3 + \alpha_2^3}. \tag{2.42}$$

Kollar et al. (2005) used these analytical results in a comprehensive model for the collision and coalescence of drops and determined the effects of these processes on the droplet size distributions. They concluded that the distribution of sizes of drops is affected significantly, not only by mass transfer processes, such as evaporation and condensation, but also by coalescence.

Of the computational schemes that may handle the coalescence of drops, the front/boundary-tracking method, which is described in more detail in Sect. 2.5.2 (Unverdi and Tryggvason 1992), has been developed to include surface tension forces and has been used to track bubbles and drops in viscous fluids. Nobari et al. (1996) also used this method to model the axisymmetric collisions of drops. Their computational results showed that the two drops deform significantly upon impact, and their fronts become flat. A very thin layer of the viscous interstitial fluid was retained between the two drops by the computational scheme, which did not have enough time to drain during the collision process. The presence of this fluid layer always caused the eventual rebounding of the drops. Coalescence in this computational scheme occurred only when the interstitial fluid layer was artificially drained using a specified condition in the numerical algorithm. The timing of the drainage process and the conditions under which the drainage of the film is applied take the place of the closure equations for the numerical method.

The study by Nobari et al. (1996) suggests that any assumptions made for the drainage of the interstitial fluid in all the numerical methods are crucial for the eventual coalescence or rebound of drops in a viscous fluid. Since for head-on collisions with significant pre-collision momentum, the minimum gap between the drops is composed of only a few molecular layers, computations at the molecular level may be needed to accurately determine the mechanics of the film drainage and the coalescence process of drops. This imposes the problem of modeling at the molecular and the continuum scales simultaneously, which is a rather challenging but not insurmountable task.

2.4.8 Heat Transfer During Collisions

The collision time is very short for all types of particles (this includes solid particles as well as drops). In addition, the area of contact between particles during the collision process is also very small, for the conduction through that area to be significant. For this reason, the heat transfer between particles during an entire collision process is negligible in comparison to the heat exchanged between particles and fluid. The fluid and the particles continue to exchange energy during the collision processes. For a spherical particle, the rate of heat transfer is given in terms of the convective heat transfer coefficient, h_c, and the temperature difference as follows:

$$\dot{Q} = 4\pi\alpha^2 h_c \left(T_p - T_f\right). \tag{2.43}$$

The dependence of the convective heat transfer coefficient, h_c, on the proximity of other particles, or during the physical collision process when deformation occurs, has not been thoroughly investigated. Given that the duration of the collisions is very short in comparison to the thermal timescale of the particles, τ_{th}, the effects of the collisions on h_c (and by extent on Nu) are typically neglected. Hence, the closure equations for Nu, presented in Sect. 1.4, may be used during collisions. In general, the effect of interparticle collisions or particle collisions with walls influence primarily the velocity of the particles, and any effects on the energy exchange come through the dependence of h_c (or Nu) on the particle velocity. In the case of drop coalescence, the resulting drop that is introduced in the computational scheme after the coalescence continuously exchanges mass and heat with the carrier fluid according to the correlations presented in Sect. 1.4.

2.5 The Treatment of Particle Boundaries

Numerical computations are performed on the nodes of the numerical grid, which typically follow a geometric pattern. Solid and fluid boundaries of particles do not necessarily coincide with these nodes. Eulerian (two-fluid) models do not model the flow and energy exchange of individual particles and do not need to account for particle boundaries. On the other hand, DNS and similar models, which account for individual particles, must accurately describe the surface of the particles that are tracked. Among the techniques that have been used for the description of the surface of the particles are the following.

2.5.1 Body-Fitted Coordinates

In the Body-Fitted Coordinates (BFC) method, the numerical grid is constructed according to the shape of the particles. Spherical particles are more easily fitted to a spherical coordinate system, but particles of other shapes may also be fitted. Alternatively, a coordinate transformation may be made (Thompson et al. 1982)

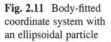

Fig. 2.11 Body-fitted
coordinate system with
an ellipsoidal particle

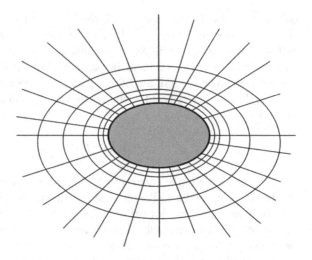

to fit the boundary of a particle to a coordinate system. If the particle shape is complex or irregular, this transformation may be accomplished numerically. Figure 2.11 shows an ellipsoidal particle fitted to a simple ellipsoidal system of coordinates. The no-slip boundary condition may be easily applied to the nodes, which describe the surface of this particle, by defining the velocity at these nodes to be equal to zero. Because the numerical grid needs to be denser close to the solid surface, oftentimes a logarithmic equation is used for the spacing of the grid nodes (Feng and Michaelides 2001). Any robust numerical scheme, such as finite difference or finite elements/volumes, may be used for the solution of the governing equations.

The BFC method is simple and ideal for the modeling of stationary, single particles, where the center of the coordinate system may coincide with the center of the particles. However, this method becomes cumbersome when several particles that are moving around are included in the computational domain. Some authors have counteracted this difficulty, by re-meshing after every time step, but this becomes computationally expensive (Patankar et al. 2001). Also, when particles are in close proximity to each other, a great deal of nodes is required to be used for the resolution of the interstitial fluid. This increases significantly the amount of computational resources that are necessary for the application of the method.

With boundary-fitted coordinates, typically, the equation of motion for the particles is solved first, and, secondly, the energy equation is computed to yield the heat transfer between the particles and the fluid. In the case of BFC, both the momentum and the energy equations for particles may be solved on the same numerical grid. Among the recent studies that used the BFC method are McKenna et al. (1999) who studied the heat transfer from catalyst spheres, Nijemeisland and Dixon (2004) who investigated the heat transfer in a fixed bed of spheres, and Gan et al. (2003) who simulated the sedimentation of solid particles with thermal convection. The last study used the arbitrary Lagrangian–Eulerian (ALE) finite element method (FEM).

The inherent disadvantage of re-meshing and ensuring for an adequate number of interparticle nodes has motivated researchers to use non-fitted methods, such as the Lattice Boltzmann or the Immersed Boundary Methods, which are examined in Sects. 2.5.3 and 2.5.4.

2.5.2 The Front-Tracking Method

The introduction of the Front-Tracking Method (FTM) for the solution of particulate flow problems started with the work of Marshall (1986), who solved the Stefan problem using this method. He formulated the Stefan problem as an ordinary differential equation, initial-value problem for the moving boundary coupled with a parabolic partial differential equation for the temperature field. The numerical calculations gave excellent results for the one-dimensional propagation of the solid front with straight and curved moving boundaries. Unverdi and Tryggvason (1992) extended the FTM to particulate and bubbly flows in viscous fluids.

The FTM avoids an implicit interface description within the carrier fluid domain. Such a description requires grid reconstruction at each time step to estimate the position and slope of the interface. Instead, the method uses a different grid on the interface surface, which is distinct from the grid of the flow domain. The second grid requires restructuring because the interface moves and deforms continuously as the calculations progress in time. Unverdi and Tryggvason (1992) discretized the carrier flow field by a finite difference approximation on a stationary cubical grid and used a two-dimensional triangular grid for the interface. The interface points are sometimes called the "marker points." Given the front location, the Eulerian marker function field as well as the corresponding density, viscosity, and force due to surface tension are determined and are used to calculate the solution, from which the Lagrangian points of the interface are advected. The FTM introduces a natural way to accommodate surface tension effects and other forces that determine the surface deformation. The method also keeps the density and viscosity stratification sharp.

An advantage of the FTM is that the unit outward normal on the interface may be directly obtained from the Lagrangian surface/front grid. This is accomplished by relating the grid gradient to a sum of the projections of the Lagrangian grid points, which define the interface:

$$\nabla F_Z(x_i) = \sum_{k=1}^{N_i} Z\vec{n}_j \delta A_j, \qquad (2.44)$$

where F_Z is the function that defines the front at the Eulerian grid points x_i and n_j is the normal vector to the elemental area δA_j of the front. This procedure results in the solution of a Poisson equation that solves for the "marker function," which defines the marker points:

$$\nabla^2 F_Z = \vec{\nabla} \cdot \left(\vec{\nabla} F_Z \right). \qquad (2.45)$$

Once the marker function and the marker points are known at a given time step, the density and effective viscosity of the Eulerian field are computed, and the computations for the carrier fluid and the interface proceed to the next step. Because volume is not explicitly preserved, renormalization is required to ensure the volume conservation at the interface. Tryggvason et al. (2001) extended the FTM and performed DNS of multiphase flows. They discussed the problem of the moving interface as well as the transfer of information between the moving front grid and the fixed Eulerian grid. They also gave examples of the application of the FTM to homogeneous bubbly flows, atomization, flows with variable surface tension, solidification, and boiling.

2.5.3 The Lattice Boltzmann Method

The Lattice Boltzmann Method (LBM) was developed in the 1990s and is based on statistical mechanics (Frisch et al. 1986, 1987). The flow field is modeled by a system of nodes (fluid particles), typically in a square or cubic arrangement. A distribution function, $f_i(x,t)$, and its evolution, $f_i(x+u_i\Delta t,t+\Delta t)$ which describes the interaction and evolution of the fluid nodes, is defined as follows:

$$f_i(\vec{x}+\vec{u}_i\Delta t, t+\Delta t) = f_i(\vec{x}, t) - \frac{f_i(\vec{x}, t) - f_i^{eq}(\vec{x}, t)}{\tau},$$ (2.46)

where $f_i^{eq}(x,t)$ is the well-defined equilibrium state of the distribution function and τ is the dimensionless relaxation time. When the latter is defined in terms of the dimensionless viscosity as

$$v^* = (2\tau - 1)/6,$$ (2.47)

it has been proven that the LBM method models a viscous fluid with kinematic viscosity, v, and that the computational error from this modeling is related to the characteristic speed of the flow, U_c; the grid timescale, Δt; and the grid spacing, Δx, through a computational Mach number, Ma. The latter is defined as

$$Ma \equiv \frac{U_c}{\frac{\Delta t}{\Delta x}} = \frac{U_c \Delta x}{v}\left(\frac{2\tau - 1}{6}\right),$$ (2.48)

where v is the actual viscosity of the fluid. When $Ma \ll 1$, the LBM describes accurately the viscous flow (Ladd 1994a). The equilibrium distribution function for a viscous fluid is given by the expression:

$$f_i^{eq}(\vec{x}, t) = \rho w_i\left[1 + 3\vec{e}_i \cdot \vec{u} + \frac{9}{2}(\vec{e}_i \cdot \vec{u})^2 - \frac{3}{2}\vec{u} \cdot \vec{u}\right].$$ (2.49)

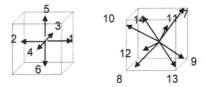

Fig. 2.12 The 14 interactions of the node at the center of the cube with nodes at the vertices and faces of the cube. The 15th interaction is the null vector

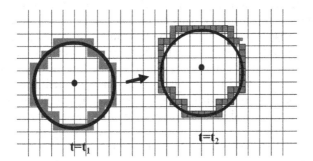

Fig. 2.13 The circular particle, in *black color*, and its surface nodes, in *gray*, at two different times

The weights w_i are well-defined constants in two and three dimensions, and the unit vectors, $\vec{e}_i$, define the ways of interaction of the fluid particles/nodes with their neighboring nodes. In the LBM method, a node is linked only to its surrounding nodes. Figure 2.12 shows the interaction links of the central node in the cube. Interactions in 15 directions are allowed, which means that the central node interacts with other nodes at the centers of the eight vertices of the cube and with the centers of the six faces of the cube. The 15th interaction is covered by the null vector, which implies that the fluid in the central node is at rest, that is, the node interacts with itself only.

In order to describe a surface, Ladd (1994a, b) and Ladd and Verberg (2001) introduced the *bounce-back rule*, which defines a surface in the LBM: According to this rule, the particle surface is represented by the so-called *boundary nodes*. The boundary nodes are a set of the midpoints of the links between two fixed grid nodes. One of the boundary nodes is within the fluid domain, and the other is within the domain of the solid particle. Figure 2.13 shows the surface boundary nodes, in gray color, for a circular particle, in solid black color, within a rectangular grid at two different times. The interactions at the boundary nodes are prescribed in a way to fulfill the zero penetration and the no-slip condition at this boundary (Ladd 1994b). The simplest way to achieve the boundary conditions is to reflect (bounce back) all the interactions and fluid elements on the boundary surface. This condition is satisfied if all the fluid elements that are directed from the fluid domain to the surface are canceled by the corresponding fluid elements emanating from the interior of the solid domain and are directed to the surface of the particle:

$$f_i^{\text{in}}(\vec{x}) = f_i^{\text{out}}(\vec{x}). \tag{2.50}$$

The application of this rather simple condition at the boundary nodes ensures the complete reflection of the fluid elements on the "surface" and the de facto application of the no-penetration and no-slip conditions. If necessary, it is possible to modify the bounce-back rule and to allow partial interaction at the interface, so that partial penetration and partial slip are allowed, as in the case of a porous boundary (Walsh et al. 2009).

The definition of particle boundaries introduces significant problems with the use of the LBM: At first, the numerical scheme makes it necessary to use a large number of lattice grids for every particle in the flow field if the physical boundaries are to be represented accurately. This necessitates a very dense grid for particles with irregular shapes. Secondly, the finite number of boundary nodes makes necessary the stepwise representation of the particle boundary. This causes fluctuations on the computation of the hydrodynamic force acting on the particle and limits the ability of the LBM to solve particle–fluid interaction problems at high Reynolds numbers. Thirdly, when a particle moves, its computational boundary changes and may vary significantly between time steps. This is apparent in Fig. 2.12, which shows that the "particle" as described by its own surface, in gray, has changed shape between the times t_1 and t_2. The surface modification after each time step causes fluctuations in the computation of forces and velocities of the particle. Several authors have used the matching of empirical, closure expressions for the drag coefficient of particles to make the shape transition smoother and to ensure that the actual computational scheme does not become unstable.

Another, but rather minor, problem associated with the application of the LBM is that the bounce-back rule treats the particle–fluid interaction only at the surface of the particle. The interior of the particle domain remains "fluid" during the computations. Thus, the rigid body motion in the interior of the particle is not a priori enforced. The problem that is actually solved by the LBM is the interaction between a fluid and a solid shell, which has a similar boundary as the particle and carries the entire mass of the particle. The contribution of the particle interior to the particle motion and fluid–particle interactions is ignored. It is fortuitous that the effect of the interior fluid is not significant in the hydrodynamic interactions of the solid particles. This was proven by Ladd and Verberg (2001) who showed that accurate computations may be carried out with or without considering the interior fluid.

2.5.4   The Immersed Boundary Method

The Immersed Boundary Method (IBM) was first developed by Peskin (1977) to model the motion of the moving boundary of the human heart. Fogelson and Peskin (1988) have showed that this method could also be employed to simulate flows with suspended, deformable, or rigid particles. Glowinski et al. (2001) assisted in the

Fig. 2.14 The four stages in the conceptual development of the IBM

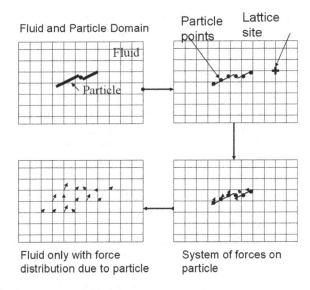

Fluid only with force distribution due to particle

System of forces on particle

development of the IBM by using Lagrange multipliers and the fictitious domain method (FDM) to enforce the no-slip boundary condition at the fluid–particle interface. Feng and Michaelides (2004) combined the IBM and the LBM by computing the force density through a penalty method in the simulations of particulate flows.

The IBM uses a fixed Cartesian mesh for the fluid, which is composed of Eulerian nodes. For the solid boundaries that are immersed in the fluid, the IBM uses a set of Lagrangian boundary nodes, which are advected by the fluid–particle interactions. It may be said that the IBM uses two computational domains: one for the fluid and one for the particulate phase. The interactions between the two domains emanate from the application of a suitable system of forces on the Eulerian fluid domain.

Feng and Michaelides (2005) extended the IBM and developed the *Proteus* numerical code, which incorporates a fictitious domain method and a direct forcing scheme to model the flow of very large numbers of particles in two and three dimensions. Shortly afterward, Uhlmann (2005) independently developed a similar numerical method. The main advantage of these two methods is that the force term is not obtained by a feedback mechanism but by a direct numerical approach. The final result on the computations is that oscillations due to the fixed grid are suppressed because the methods have the ability to smoothly transfer variables between the Lagrangian and Eulerian domains. Another advantage of the IBM is the direct and explicit formulation of the fluid–solid interaction force. Because of this, the IBM with direct forcing produces weaker artificial oscillatory transient particle forces than other methods and has resulted in higher computational efficiency and accuracy compared to the indirect methods.

Figure 2.14 depicts the conceptual design of the IBM, with the four diagrams showing the stages of the development of the numerical scheme. The arrows connecting the four diagrams follow the development of the method. A particle, which may be rigid or deformable, is in the domain of the carrier fluid, and a Eulerian numerical grid is applied to the entire fluid–particle domain. In the first diagram of the figure, the particle appears as a fiber that may be deformable or rigid.

In the second stage of the method, the surface of the particle is discretized by a number of surface points. The number of the points chosen must be sufficient to describe the surface of the particle within the required degree of accuracy. It must be noted that a more accurate representation of deformable particles requires a higher number of surface points as well as the accurate description of the forces that resist the deformation, e.g., surface tension forces or chemical bonds. A system of springs is chosen to connect each surface point with its neighbors, and the stiffness (spring constant) of the springs will determine the deformation of the surface. For example, very high stiffness will result in rigid particles, while low stiffness results in easily deformable particles.

The third stage in the application of the method determines the system of the hydrodynamic forces acting on the surface of the particle and applies them on the points that represent the surface of the particle. Since the surface points do not coincide with the fluid lattice points, this system of forces is appropriately transposed on the fluid lattice points that are neighbors to the particle surface points.

The transposition of the forces is the fourth stage of the IBM and is shown schematically in the fourth diagram of Fig. 2.14. The net effect, which is apparent between stages 1 and 4 and which defines the IBM, is the substitution of the surface of the particle by an equivalent system of forces, which has the same effect on the fluid as the surface of the particle. Hence, the Navier–Stokes equations for the fluid domain include an additional force, $\vec{f}$, which is due to the presence of the particle:

$$\rho_f\left(\frac{\partial \vec{u}}{\partial t} + \vec{u} \cdot \nabla \vec{u}\right) = \mu \nabla^2 \vec{u} - \vec{\nabla}P + \vec{f}. \tag{2.51}$$

This force vanishes at the fluid lattice sites that do not neighbor the particle surface points. The no-slip boundary condition at the interface is automatically satisfied by enforcing the velocity at all boundaries to be equal to the velocity of the fluid at the same location:

$$\frac{\partial \vec{X}(s,t)}{\partial t} = \vec{u}(X(s,t),t), \tag{2.52}$$

where s is the parameter that represents the points on the surface of the particle, and $x = X(s,t)$ is the representation of the particle surface function in the Eulerian domain. Surface slip and penetrating conditions may also be prescribed by modifying Eq. (2.52).

2.5.5   Application of the IBM to Heat Transfer

Particulate heat transfer may also be studied directly using the IBM. For the computation of the heat transfer, Yu et al. (2006) employed the fictitious domain method to study two-dimensional particulate flow with heat convection. They used Lagrangian multipliers to resolve the heat interactions between the fluid and particles. Kim and Choi (2004) used a version of the IBM to study heat transfer problems with stationary particles and complex geometries. Also Pacheco et al. (2005) presented an IBM based on the finite-volume method to study the heat transfer and fluid flow problems with non-staggered grids.

Feng and Michaelides (2008, 2009) extended the IBM in a straightforward and direct way to apply to the energy equation. They introduced an approach that utilizes the main premise of the IBM for the solution of the energy interaction between particles and fluid. According to this approach, the modified momentum and energy equations are solved only on the Eulerian grid. This provides a simplification for the overall numerical technique and requires significantly lower computational resources and CPU time. They postulated that the surface of particles, which exchange thermal energy with the fluid, may be substituted by a system of discretized heat sources and sinks. The net effect of this system of heat sources and sinks on the fluid is to exchange the same amount of heat between the particles and the fluid. Thus, the energy equation for the fluid is modified to include the heat sources that represent the particles as follows:

$$\rho_f c_f \frac{\partial T}{\partial t} + \rho_f c_f \vec{u} \cdot \nabla T = k_f \nabla^2 T + q_{int} + q_{sur}, \qquad (2.53)$$

where q_{sur} is the heat that is exchanged due to the heat sources and sinks at the surface of the particle and q_{int} represents any other internal heat sources the fluid may have. The former is the result of the application of the IBM on the energy equation for the fluid–particle system. The heat sources and sinks may be placed on the same boundary nodes where the IBM forces act by a similar transposition technique. Hence, the same point discretization scheme may be used for the momentum and the energy equations, a fact that significantly simplifies the computational method and accelerates the computations.

The IBM is ideally suited for the simulation of the effects of deformed immersed boundaries and has been widely used in biological fluid dynamics. The method is robust and may handle very large numbers of interacting and deformable particles, such as biological cells. The addition of the capability to model the energy and mass exchange between the fluid and the particles makes it ideally suited for applications where momentum, mass, and energy exchange are important for the modeling of discrete particles, such as blood cells, drug delivery, and fluidized bed reactors.

Bibliography

Ashgriz N, Poo JY (1990) Coalescence and separation in binary collisions of liquid drops. J Fluid Mech 221:183–204

Boure JA, Delhaye J-M (1982) General equations and two-phase modeling. In: Gad H (ed) Handbook of multiphase systems. Hemisphere, New York

Chen S, Doolen GD (1998) Lattice Boltzmann method for fluid flows. Annu Rev Fluid Mech 30:329–364

Crowe CT, Sommerfeld M, Tsuji Y (1998) Multiphase flows with droplets and particles. CRC, Boca Raton, FL

Cundall PA, Strack OD (1979) A discrete numerical model for granular assemblies. Geotechnique 29:47–54

Davis A, Michaelides EE, Feng Z-G (2011) Particle velocity near vertical boundaries – a source of uncertainty in two-fluid models. Powder Technol. doi:10.1016/j.powtec.2011.09.031

Delhaye J-M (1981) Basic equations for two-phase modeling. In: Bergles AE, Collier JG, Delhaye J-M, Hewitt GF, Mayinger F (eds) Two-phase flow and heat transfer in the power and process industries. Hemisphere, New York

Ding J, Gidaspaw D (1990) A bubbling fluidization model using kinetic theory of granular flow. AIChE J 36:523–538

Estrade JP, Carentz H, Laverne G, Biscos Y (1999) Experimental investigation of dynamic binary collision of ethanol droplets—a model for droplet coalescence and bouncing. Int J Heat Fluid Flow 20:486–491

Feng Z-G, Michaelides EE (2001) Heat and mass transfer coefficients of viscous spheres. Int J Heat Mass Transf 44:4445–4454

Feng Z-G, Michaelides EE (2004) An immersed boundary method combined with lattice Boltzmann method for solving fluid and particles interaction problems. J Comput Phys 195:602–628

Feng Z-G, Michaelides EE (2005) Proteus—a direct forcing method in the simulation of particulate flows. J Comput Phys 202:20–51

Feng Z-G, Michaelides EE (2008) Inclusion of heat transfer computations for particle laden flows. Phys Fluids 20:1–10

Feng Z-G, Michaelides EE (2009) Heat transfer in particulate flows with direct numerical simulation (DNS). Int J Heat Mass Transf 52:777–786

Fogelson AL, Peskin CS (1988) A fast numerical method for solving the three-dimensional Stokes equation in the presence of suspended particles. J Comput Phys 79:50–69

Frank T, Schade KP, Petrak D (1993) Numerical simulation and experimental investigation of gas-solid two phase flow in a horizontal channel. Int J Multiphase Flow 19:187–204

Frisch U, Hasslacher B, Pomeau Y (1986) Lattice-gas automata for the Navier-Stokes equations. Phys Rev Lett 56:1505

Frisch U, D'Humiéres D, Hasslacher B, Lallemand P, Pomeau Y, Rivert JP (1987) Lattice-gas hydrodynamics in two and three dimensions. Complex Syst 1:649–707

Gan H, Chang JZ, Howard HH (2003) Direct numerical simulation of the sedimentation of solid particles with thermal convection. J Fluid Mech 481:385–411

Glowinski R, Pan T-W, Hesla TI, Joseph DD, Periaux J (2001) A fictitious domain approach to the direct numerical simulation of incompressible viscous flow past moving rigid bodies: application to particulate flow. J Comput Phys 169:363–426

Gosman AD, Ioannides E (1983) Aspects of computer simulation of liquid-fueled reactors. Energy 7:482–490

Hinze JO (1975) Turbulence. McGraw-Hill, New York

Höfler K, Schwarzer S (2000) Navier-Stokes simulation with constraint forces: finite-difference method for particle-laden flows and complex geometries. Phys Rev E 61:7146–7160

Ishii M (1975) Thermo-fluid dynamic theory of two-phase flows. Eyrolles, Paris

Ishii M (1990) Two-fluid model for two-phase flow. In: Hewitt GF, Delhaye J-M, Zuber N (eds) Multiphase science and technology, vol 5. Hemisphere, New York, pp 1–63

Jenkins JT, Richman MW (1985) Grad's 13-moment system for a dense gas of inelastic spheres. Arch Ration Mech Anal 87:355–377

Kartushinsky A, Michaelides EE (2004) An analytical approach for the closure equations of gas-solid flows with inter-particle collisions. Int J Multiphase Flow 30:159–180

Kim J, Choi H (2004) An immersed-boundary finite-volume method for simulations of heat transfer in complex geometries. Korean Soc Mech Eng Int J 18(6):1026–1035

Kollar LE, Farzaneh M, Karev AR (2005) Modeling droplet collision and coalescence in an icing wind tunnel and the influence of these processes on droplet size distribution. Int J Multiphase Flow 31:69–92

Ladd AJC (1994a) Numerical simulations of particulate suspensions via a discretized Boltzmann equation. Part 1. Theoretical foundation. J Fluid Mech 271:285–310

Ladd AJC (1994b) Numerical simulations of particulate suspensions in a discretized Boltzmann equation Part II. Numerical results. J Fluid Mech 271:311–339

Ladd AJC, Verberg R (2001) Lattice-Boltzmann simulation of particle-fluid suspensions. J Stat Phys 104:1191–1251

Manga M, Stone HA (1993) Buoyancy-driven interactions between two deformable viscous drops. J Fluid Mech 256:647–683

Manga M, Stone HA (1995) Low Reynolds number motion of bubbles, drops and rigid spheres through fluid-fluid interfaces. J Fluid Mech 287:279–298

Marshall G (1986) A front tracking method for one-dimensional moving boundary problems. SIAM J Sci Stat Comput 7:252–263

McKenna TF, Spitz R, Cokljat D (1999) Heat transfer from catalysts with computational fluid dynamics. AIChE J 45:2392–2410

Michaelides EE (2003) Introduction and basic equations for multiphase flow. In: Buchlin J-M (ed) von Karman institute lecture series, 19–23 May 2003, Brussels, Belgium

Michaelides EE, Li L, Lasek A (1992) The effect of turbulence on the phase change of droplets and particles under non-equilibrium conditions. Int J Heat Mass Transf 34:2069–2076

Morioka S, Nakajima T (1987) Modeling of gas and solid particles two-phase flow and application to fluidized bed. J Theor Appl Mech 6:77–88

Mostafa AA, Elghobashi SE (1985) A two-equation turbulence model for jet flows laden with vaporizing droplets. Int J Multiphase Flow 11:515–533

Nijemeisland M, Dixon AG (2004) CFD study of fluid flow and wall heat transfer in a fixed bed of spheres. AIChE J 50:906–921

Nobari MRH, Jan YJ, Tryggvason G (1996) Head on collisions of drops – a numerical investigation. Phys Fluids 8:29–42

Orme M (1997) Experiments on droplet collisions bounce, coalescence and disruption. Prog Energy Combust Sci 23:65–79

Pacheco JR, Pacheco-Vega A, Rodic T, Peck RE (2005) Numerical simulations of heat transfer and fluid problems using an immersed-boundary finite-volume method on non-staggered grids. Numer Heat Transf B 48:1–24

Patankar N, Ko T, Choi HG, Joseph DD (2001) A correlation for the lift-off of many particles in plane Poiseuille flows of Newtonian fluids. J Fluid Mech 445:55–76

Peskin CS (1977) Numerical analysis of blood flow in the heart. J Comput Phys 25:220–252

Peskin CS (2002) The immersed boundary method. Acta Numer 11:479–517

Piomelli U, Balaras E (2002) Wall-layer models for large-eddy simulations. Annu Rev Fluid Mech 34:349–374

Qian J, Law CK (1997) Regimes of coalescence and separation in droplet collisions. J Fluid Mech 331:59–80

Reeks MW (1991) On a kinetic equation for the transport of particles in turbulent flows. Phys Fluids 3:446–456

Reeks MW, Simonin O (2006) PDF models. In: Multiphase flow handbook, sect. 13.4.3. CRC, Boca Raton, FL, pp 13-89–13-113

Sakiz M, Simonin O (1999) Development and validation of continuum particle wall boundary conditions using Lagrangian simulations of a vertical gas-solids channel flow. In: Proceedings of 3rd ASME-JSME joint fluids engineering conference, San Francisco, CA

Simonin O (2001) Statistical and continuum modeling of turbulent reactive particulate flows- Part I: Theoretical derivation of dispersed phase Eulerian modeling from probability density function kinetic equation. In: Buchlin J-M (ed) von Karman institute lecture series. Buchlin, Brussels

Sirignano WA (1999) Fluid dynamics and transport of droplets and sprays. Cambridge University Press, Cambridge

Smagorinsky J (1963) General circulation experiments with the primitive equations, I. The basic experiment. Mon Weather Rev 91:99–152

Sommerfeld M (2003) Analysis of collision effects for turbulent gas-particle flow in a horizontal channel: Part I. Particle transport. Int J Multiphase Flow 29:675–699

Sommerfeld M, Huber N (1999) Experimental analysis and modelling of particle-wall collisions. Int J Multiphase Flow 25:1457–1489

Taniere A, Khalij M, Oesterle B (2004) Focus on the disperse phase boundary conditions at the wall for irregular particle bouncing. Int J Multiphase Flow 30:675–699

Thompson JF, Warsi ZUA, Mastin CW (1982) Boundary-fitted coordinate systems for numerical solution of partial differential equations—a review. J Comput Phys 47:1–108

Tryggvason G, Bunner B, Esmaeeli A, Juric D, Al-Rawahi N, Tauber W, Han J, Nas S, Jan Y-J (2001) A front-tracking method for the computations of multiphase flow. J Comput Phys 169:708–759

Tsuji Y, Kawaguchi T, Tanaka T (1993) Discrete particle simulation in a two dimensional fluidized bed. Powder Technol 77:79–86

Uhlmann M (2005) An immersed boundary method with direct forcing for the simulation of particulate flows. J Comput Phys 209:448–476

Unverdi SO, Tryggvason G (1992) A front-tracking method for viscous, incompressible, multi-fluid flows. J Comput Phys 100:25–37

Wallis GB (1963) Some hydrodynamic aspects of two-phase flow and boiling. In: International developments in heat transfer. ASME, New York

Walsh STC, Burnwinkle H, Saar MO (2009) A new partial-bounce back lattice-Boltzmann method for fluid flow through heterogeneous media. Comput Geosci 35:1186–1193

Warsi ZUA (1993) Fluid dynamics: theoretical and computational approaches. CRC, Boca Raton, FL

Yu Z, Shao X, Wachs A (2006) A fictitious domain method for particulate flows with heat transfer. J Comput Phys 217:424–452

Zaichik LI, Vinberg AA (1991) Modeling of particle dynamics and heat transfer in turbulent flows using equations for first and second moments of velocity and temperature fluctuations. In: Proceedings of 8th international symposium on turbulent shear flows, Munich, FRG, vol 1, pp 1021–1026

Chapter 3
Fluidized Bed Reactors

Keywords Fluidized beds • Distributors • Fluidization velocity • Catalytic crackers • Catalytic synthesis • Combustors • Gasifiers • Numerical codes

The name "fluidized bed reactors" (FBR) is a generic term and encompasses a variety of engineering systems including chemical reactors, combustors, gasifiers, calcifiers, driers, etc. The common characteristic of the FBR class of engineering systems is the presence of the solid particles that are carried by the fluid and exist in a fluidized state. The fluid effectively lifts the solid particles and carries them to different parts of the FBR, where the fluid velocity is lower, e.g., close to the walls of the FBR. There, the fluid drag is insufficient to keep the particles suspended, and the particles fall to parts of the system where the fluid velocity is high enough to lift them again. This circular particle motion enhances any flow instabilities, such as vortices and turbulence, and results in very high levels of mixing for both the particles and the carrier fluid. The fluid–particle interactions typically include momentum exchange, mass exchange (reactions), and energy/heat exchange. The presence and movement of the solid particles in a fluidized state add to the agitation within the FBR and, thus, enhances all fluid–particle interactions as well as the mass, energy, and momentum exchanges. Figure 3.1 depicts two generic types of FBRs: the flat bed and the spouted bed reactors. The general flow pattern in both reactors is that particles rise to the top of the reactors primarily through the center, move laterally to the sides, and settle close to the walls. In the spouted FBR, the settling particles also move laterally in the spout area toward the center of the FBR. The broad arrows in the figure indicate the general flow patterns. It must be noted that, typically, the fluid outlets include a cyclone separator and/or a filter, which are not shown in the figure. These devices separate any entrained particles from the carrier fluid at the exits of the FBR. The separated particles are typically fed back to the FBR.

E.E. (Stathis) Michaelides, *Heat and Mass Transfer* 89
in Particulate Suspensions, SpringerBriefs in Applied Sciences and Technology,
DOI 10.1007/978-1-4614-5854-8_3, © Springer Science+Business Media New York 2013

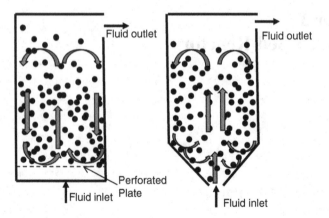

Fig. 3.1 Flow mixing and general particle motion patterns in flat bed and spouted FBRs

Some of the inherent advantages of the FBRs compared to other reactor systems are:

1. Significantly higher mixing of both solids and fluid. Because the particles flow and move inside the FBR, they facilitate the mixing of the fluid. The effect of the higher mixing in the fluid is to develop more uniform concentration and temperature fields. This avoids the creation of "hot spots" or "cold spots" in the reactor, which invariably result in product deterioration and reactor malfunction.
2. Because of the high mixing, the FBRs also exhibit significantly higher heat and mass transfer coefficients to enclosed surfaces. This results in less surface requirements for a given rate of heat transfer.
3. The vigorous mixing causes the minimization of the temperature and concentration gradients. In particular, because of this mixing, the radial and azimuthal gradients may be almost eliminated. This invariably produces a more uniform and higher quality product. The longitudinal gradients are typically taken into account in the optimum design of the reactor and are sustained for the smooth operation of the reactor and the production of quality products.
4. In comparison to the packed bed solid-phase reactors, the flow of the solids in the FBRs causes a more uniform particle mixing and more uniform reactions. The superior mixing of the solid particles produces a uniform product usually of superior quality, which is more difficult to achieve in other types of reactors.
5. Since the volumetric heat capacity of solids (in J/m^3) is much higher than that of the fluid ($c_s\rho_s \gg c_f\rho_f$), the solids provide a very effective way to add or remove heat out of the reactor and to maintain a constant temperature for both endothermic and exothermic reactions.
6. The control of the reactions in a FBR has been significantly improved. As more research is performed and the engineering experience with FBRs is increasing, it is becoming easier to design better-performing FBRs and control the reactions to produce the desired products with satisfactory quality and yield.

7. Results from continued research on solid–fluid phase interactions and better computer modeling and simulations are used to produce better and more efficient designs that serve increasingly more applications. This has helped create an expanding market for the FBRs with significant financial advantages for their operators.

8. Perhaps the most important advantage of the FBRs is the capability to operate continuously and not make use of batch processes. The FBR is an open thermo-dynamic system, with inlets that may be supplied continuously with reactants and outlets that exhaust continuously the products. The continuous and uniform operation of the FBRs enables a more efficient and uniform production process for all their products because the start-up and finish conditions are eliminated. There is no delay in the production process for the removal of the products and the supply of the reactants.

Inherent disadvantages of the FBR systems include:

1. More expensive design for the overall system because the FBRs include several moving parts.

2. The requirement that the fluid causes suspension of the solid material imposes higher fluid velocities and higher fluid pressures than in packed bed reactors. In the case of FBRs operating with a gas, this may add significantly to the power requirements for the compression of the gas.

3. Related to the above is the stoppage of the equipment if the higher fluidization pressure is lost, e.g., because of a blower or compressor malfunction. This may result in product degradation or unexpected and undesirable reactions. Such a malfunction may be avoided by better design and maintenance schedule of the equipment.

4. In comparison to the packed bed reactors, where the solid particles are station-ary, the FBRs have higher reactor volumes because the solid phase expands.

5. The higher gas velocities, which are inherent in FBRs, result in the entrainment of a larger amount of solid particles, especially fines, which may be carried by the fluid outside the reactor. This causes the unnecessary waste of reactants, which may become expensive and in some cases environmentally undesirable. Better reactor design and suitable entrainment reducing technologies may be utilized to reduce the solids entrainment problem. For example, one or two cyclone separators with a filter for ultrafine particles may be used at the outlet of the FBR.

6. They represent relatively new technology. All the aspects of the operation of the FBRs are not fully understood. Despite the enormous research efforts on FBRs since the 1970s, several aspects of their operation are not well understood and are poorly modeled. A concerted effort has been under way (see Sect. 3.5) to better model FBR technology and to remove some of the design uncertainties associated with this technology. This worldwide research effort has started bearing results that make newly designed FBRs better and more efficient to operate.

7. Because of the lack of complete understanding of the FBR technology, pilot plants for new processes or products are necessary to be built. Since the FBRs have multiple length-scales that are important, the scale-up of pilot plants to fully producing facilities is undocumented in the open literature and uncertain. Oftentimes, the results of full-scale facilities do not reflect what was experienced in the pilot plant trial.

8. Related to the above and the high economic value of FBRs, a great deal of engineering experience with FBRs is proprietary, and several parts and processes of FBRs are covered by patents. For this reason, the "best engineering practices" are not always well known and, if they are patented, may not be duplicated.

9. Particulate flow inside the reactor causes higher rates of erosion and wear on the reactor vessel. Erosion may be avoided by (a) using erosion-resistant materials, (b) better design of the flow patterns in the FBR, or (c) more frequent and expensive maintenance for the reactor vessel and piping replacement.

It appears that the inherent advantages of the FBRs outweigh the disadvantages associated with their uses. The engineering community has seen the more widespread use of FBRs in chemical engineering applications such as coal combustion, petroleum refining, calcination, and the production of chemicals. Better and more efficient FBRs continuously substitute batch reaction systems and packed reactor beds.

3.1   Types of FBRs and Air Distributors

There are two generic types of fluidized bed reactors: flat bed and spouted bed reactors. The spouted bed reactors include a spouted entrance section, where the area-averaged fluid velocity gradually decreases. Spouted reactors are currently used for specific applications in the chemical industry, while flat bed FBRs are more commonly used in large industrial applications. The advantages of the spouted bed in comparison to the flat bed FBRs are:

1. Lower pressure drop for a given depth and solids loading
2. More predictable solids and gas flow patterns
3. Lower gas flow rates and better mixing
4. Better operation with larger particles

The disadvantages of spouted bed FBRs vis-a-vis the flat bed FBRs are:

1. Lower bed-to-wall surface area, which implies lower rates of heat transfer.
2. The mass flow rate of gas is limited by the spouting area and orifice.
3. There is a maximum height, H_{max}, above which spouting does not occur. This limits the scaling of spouted reactors and the reproducibility of their operation at larger scales.
4. The spouted bed operation deteriorates when the particle size distribution becomes large because the larger particles settle in the bottom and do not fluidize.

At the bottom part of both types of FBRs, there is a *grid* or *distributor*, which distributes the fluid uniformly over the cross-sectional area of the reactor. Well-designed grids support the weight of the solids; prevent the back-flow of solids into the fluid domain, even when the flow is stopped; direct the particulates in a way that minimizes the erosion of internal equipment; and are resistant to higher temperatures, corrosion, and erosion. The function of the grid is crucial in the operation of the FBR because the uniformity of the fluid distribution and the initial formation of fluid jets or bubbles at the fluid entrance depend on the geometric characteristics of the grid, the fluid pressure field developed, and any flow characteristics behind the grid. Typical grid configurations that are used in industrial practice are:

1. *Plates* with multiple orifices, which may be horizontal or vertical. The orifices are of a variety of shapes to suit the applications.
2. *Grids* composed of grate bars or pipes.
3. *Porous plates* are common in the research laboratory but are impractical for large-scale FBRs.
4. *Screens* by themselves do not allow for good fluid distribution and are typically used in combination with other grid types.
5. *Nozzles, tuyères, and bubble caps*, which may direct the flow to be horizontal, vertical, or angled.
6. *Central openings with conical sections*, which are used in small spouted beds and in beds where the solids are supplied together with the fluid.
7. *Packed beds*, which are beds of coarser and heavier particles that do not become fluidized. The packed beds are well supported from below and often are confined at the top with a perforated plate or a screen.

3.2   Basics of the Operation of FBRs

The FBRs evolved from the *packed beds* of particles, which have been used in the chemical industry since the early nineteenth century. In packed beds, a reacting fluid, either gas or liquid, flows with relatively low velocity between the stationary particles of the bed and chemically reacts with the solids. The fluid flow in packed beds is, essentially, flow through a porous medium. The fluid is mechanically dispersed by the presence of the particles and mass, and energy exchanges occur between the fluid and the particles. However, the fluid concentration is limited by the porosity of the packed bed, particles do not move, a limited area of the particles is exposed to the flow, and chemical reactions occur at rather slow rate.

Industrial practice showed since the 1950s that, if the fluid velocity is increased, the drag force on the particles increases and, when the *fluidization velocity* is reached, a large fraction of the particles rises from the bed and is entrained in the fluid flow. When full fluidization occurs, the entrained particles are entirely supported by the fluid drag force. The latter is manifested in the rather large pressure difference between the bottom and top of the bed. The entrainment causes the upward and sideways movement of the particles in the entire bed and agitation in the fluid. This increases significantly the rates of heat and mass transfer between fluid and particles. When fluidization occurs, the fluid–particle mixture has the appearance, behavior, and many characteristics of a fluid. The mixture may be described by fluid transport properties, such as viscosity, thermal conductivity, and convective heat and mass transfer coefficients. In comparison to the packed reacting beds, FBRs have the advantages that they exhibit higher temperature and concentration uniformity, they operate at lower pressure for the fluid phase, they may accommodate a wider variety of particles and fluids, and they show superior heat and mass transfer characteristics.

It must be noted that typical FBRs, which are used in industrial processes, are built with complex flow geometries and there is a multitude of mass, energy, and momentum interactions between the fluid and the solid phases. Because of this, it is often impossible to describe the FBR operation using first principles and simple governing equations. For this reason, the literature of FBRs is filled with correlations for all the interactions between the two phases as well as between the fluid–solid mixture and the internal equipment and walls. The correlations stem from experimental data or numerical simulations, and their accuracy is limited to the range of conditions prevailing during the experiments.

3.2.1 Fluidization Regimes

The first classification of the flow regimes in FBRs was established by Geldart (1973) who studied the fluidization characteristics of particles in air and suggested the following four categories of solids:

1. Cohesive (C) particles with very small sizes that exhibit strong interparticle forces and, because of this, form clusters and aggregates. Cohesive particles fluidize poorly.
2. Aeratable (A) particles that have, in general, larger sizes and exhibit weaker interparticle forces. They fluidize readily, exhibit high mixing in a fluidized state, and deaerate slowly when the flow is interrupted. Aeratable solid materials are the most suitable materials for FBRs.
3. Bubble readily (B) particles of larger sizes with weak interparticle forces. They mix relatively well in the fluidized state.
4. Inertial (D) particles with negligible interparticle forces and dominant inertia. They fluidize readily at higher velocities, but their mixing in the fluidization state is poor because of their larger size.

Grace (1986) extended this classification to other types of gases and two-phase flow systems. It is apparent from the last two studies that the Archimedes number,

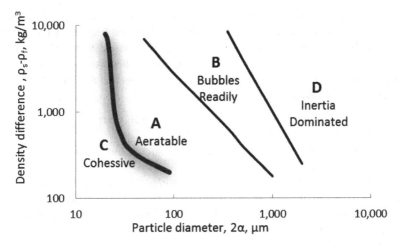

Fig. 3.2 Particle fluidization regimes in air according to Geldart (1973)

Ar, which is a measure of the ratio of the gravity/buoyancy to the viscous forces on particles, plays an important role in the fluidization process and governs the classification of particles:

$$Ar = \frac{8g\rho_f(\rho_s - \rho_s)\alpha^3}{\mu_s^2}. \tag{3.1}$$

Figure 3.2, which was reproduced from the data of Geldart (1973), shows the fluidization regimes of several types of particles in air. The boundary between the cohesive and aeratable regimes is shadowed because the cohesive forces between particles depend on the relative humidity of air. As a consequence, this boundary is not well defined. The other two boundaries are well defined and may be expressed in terms of the Archimedes number as follows (Grace 1986):

(a) For the bubbles readily to aeratable boundary, where the interparticle forces are strong but do not dominate:

$$Ar = \frac{8g\rho_f(\rho_s - \rho_s)\alpha^3}{\mu_s^2} = 10^6 \left(\frac{\rho_s - \rho_f}{\rho_s}\right)^{-1.275} \tag{3.2}$$

(b) For the inertia dominated to the bubbles readily boundary, where interparticle forces are very weak and where viscous forces and inertia dominate:

$$Ar = \frac{8g\rho_f(\rho_s - \rho_s)\alpha^3}{\mu_s^2} = 1.45 \times 10^5 \tag{3.3}$$

In most of the applications, the systems designer does not have a choice on the characteristics of the particles that will enter the fluidized bed. In such cases, the design of the flow patterns of the FBR and the processes in the FBR must be designed for the types of particles that are to be used. Whenever there is a choice of the particles in the FBR, the following particle properties and attributes are considered useful for fluidization applications:

1. Rounded, to the extent possible, with minimum sharp edges, hooks, and extensions that make particles cling together.
2. Dry enough for moisture not to make them sticky, but not too dry to induce electrostatic effects.
3. Uniform particle density. Density variations cause particle segregation.
4. Particle sizes in the range 50–1,200 μm, which implies that particles are aeratable or bubble readily. A midrange distribution of sizes is preferable to either narrow or wide distribution of sizes.
5. Resistive to attrition.
6. Midrange hardness to avoid attrition as well as erosion of equipment.

3.2.2   Minimum Fluidization Velocity

Let us consider a flat bed FBR with all the particles initially settled on the perforated *distributor plate* at the bottom of the bed. The fluid is supplied to the space below the distributor plate and the pressure rises. When the particles are settled and the bed is stationary, the pressure drop across the bed of particles is given by the so-called "Ergun's equation" (Ergun 1952) as modified by Grace (2006):

$$\frac{\Delta P_{sb}}{\Delta z} = \frac{150 \mu_f U_f \left(1 - \overline{\phi}\right)^2}{4 \alpha^2 \overline{\phi}^3 \Psi^2} + \frac{1.75 \rho_f U_f^2 \left(1 - \overline{\phi}\right)}{2 \alpha \overline{\phi}^3 \Psi}, \tag{3.4}$$

where $\overline{\phi}$ is the volume fraction of the solid particles, averaged over the solids volume defined by the vertical distance, Δz; U_f is the "superficial" fluid velocity, which is defined as the volumetric flow rate of the fluid divided by the entire cross-sectional area of the FBR; and Ψ is the shape factor of the particles, which is defined in Eq. (1.36) and influences the drag on the particles.

When the bed is at its fluidized state and the particles are suspended, simple mechanical equilibrium implies that the pressure difference between any two vertical positions balances the weight of the fluid and the solids:

$$\Delta P = \left[\left(1 - \overline{\phi}\right)\rho_f + \overline{\phi}\rho_s\right] g \Delta z. \tag{3.5}$$

At the onset of fluidization, the pressure drop across the entire bed is equal to the pressure drop given by Ergun's equation (3.4). This yields the condition for *the minimum fluidization velocity* of the FBR:

$$\frac{1.75\rho_f\left(U_f^{min}\right)^2(1-\overline{\phi})}{2\alpha\overline{\phi}^3 c} + \frac{150\mu_f U_f^{min}(1-\overline{\phi})^2}{4\alpha^2\overline{\phi}^3 c^2} = \left[(1-\overline{\phi})\rho_f + \overline{\phi}\rho_s\right]g. \quad (3.6)$$

Equation (3.6) is a quadratic equation for U_f^{min}, which may be solved to yield the values for the minimum fluidization superficial velocity U_f^{min}. It must be noted, however, that the numerical values obtained from this rather simplified procedure are subjected to an uncertainty of about 25% (Grace 2006). For this reason it is always recommended that the minimum fluidization velocity be calculated by measurements in a pilot FBR. Part of the uncertainty in predicting more accurately the velocity U_f^{min} stems from the fact that the entire bed of particles does not fluidize spontaneously. Air "bubbles" and "channels" have been observed to be formed initially in fluidized beds. The bubbles and channels allow a portion of the particles to become fluidized, but also have the effect of keeping other particles stationary in areas of the FBR where the fluid velocity is lower. Thus, under some forms of bubbling and channeling, the bed is partly stationary and partly fluidized. This implies that the actual values for the pressure difference, ΔP, are less than those predicted by Eq. (3.5).

A different correlation for the minimum fluidization velocity has been derived by Aerov and Todes (1968) who postulated that minimum fluidization occurs at $\overline{\phi} = 0.4$. Thus, Aerov and Todes (1968) used Ergun's equation for the pressure drop for $\overline{\phi} = 0.4$ and derived the following expression for the Reynolds number of particles at the minimum fluidization condition:

$$Re_{mf} = \frac{2\alpha U_f^{min}}{\nu_f} = \frac{Ar}{1400 + 5.22\sqrt{Ar}}. \quad (3.7)$$

It must be noted that the above expressions apply to flat bed FBRs, where the superficial velocity is uniform. The analysis for a spouted bed would be different because the superficial fluid velocity, U_f, is variable in the diverging part of the bed and is actually inversely proportional to the square of the distance from the bottom of the bed. A spouted bed consists of the following regions (a) a relatively dilute-flow spout, which extends from the orifice to the surface of the bed; (b) the extension of the spout to a "fountain" of particles above the surface of the bed; and (c) a dense, packed bed of particles that move slowly laterally and downward to replenish the particles that are trapped by the flow in the spout. A *minimum spouting velocity* has been experimentally determined by Mathur and Gishler (1955) as follows:

$$u_{ms} = \frac{2\alpha}{D}\left(\frac{D_{or}}{D}\right)^{1/3}\sqrt{\frac{2gH(\rho_s - \rho_f)}{\rho_f}}, \quad (3.8)$$

where D is the diameter of the bed, D_{or} is the diameter at the orifice of the spouted bed, and H is the bed height.

Bed fluidization and the minimum fluidization velocity depend very much on the type of FBR as well as the type, physical condition, and characteristics of the solid particles, e.g., density, moisture, "stickiness," and electric charges. In addition to air supply at higher pressure, industrial processes with FBRs often use supplementary methods to induce fluidization and to lower the minimum fluidization velocity. Among these methods are:

1. Pulsation of the air supply.
2. Induced vibrations at selected parts of the FBR by mechanical or acoustic means.
3. Mechanical stirring of the FBR.
4. Addition of a small percentage of particles, which are easy to fluidize. Dry powders often serve in this role.
5. Not allowing the solids to settle for long periods after they are fed to the FBR. The settling of solid particles expels the interstitial air, the particles cling together, form aggregates, and become difficult to fluidize subsequently.

3.3   Heat Transfer in FBRs

The main reason for the development of the FBRs is their superior heat and mass transfer characteristics. The continuous and constantly varying movement of both fluid and particles in the reactor creates high levels of agitation, disturbs continuously the temperature and concentration gradients in the fluid, and results in significantly higher rates of heat and mass transfer. There are two types of heat and mass transfer in FBRs:

(a) The internal heat and mass transfer between the solid particles and the interstitial gas
(b) The external heat and mass transfer to surfaces in contact with the particulate system, either plates or channels

For chemical reactors, the interest of the engineer is in the chemical products of the FBR, which depends on the internal heat and mass transfer. In the case of the combustors, the interest of the engineer is in the external heat transfer to the working fluid.[1] Hence, the designs of these two types of FBRs are optimized to ensure high product quality in the first type and complete combustion of the solids with high rate of heat transfer in the second type.

The rate of heat transfer between any two objects is defined in terms of an overall heat transfer coefficient, h : $\dot{Q} = hA(T_{fb} - T_w)$. In the case of heat transfer from

---

[1] In order to avoid unnecessary repetition, only the term "heat transfer" will be used in the rest of this chapter. It must be understood that, because of the analogy of the heat and mass transfer processes, all the mechanisms described and all the results pertain to both heat and mass transfer processes.

FBRs to an external surface, either plate of channel, there are four mechanisms or modes for the transfer of heat:

(a) Convective heat transfer between the fluid and the surface
(b) Conductive heat transfer between the solids and the surface, when the two are in contact
(c) Radiation heat transfer from the fluid to the surface
(d) Radiation heat transfer from the solids to the surface

A good approximation is to consider the four mechanisms independently, which implies that the four coefficients of heat transfer are additive. Hence, the overall heat transfer coefficient may be given as the sum of the four independent heat transfer coefficients:

$$h = h_{cf} + h_{cs} + h_{rf} + h_{rs}. \tag{3.9}$$

Because the contact area and the contact time of the solids with any surface are very short, as explained in Sect. 2.4.8, the solid-to-surface conduction may be considered negligible. In this case, the first two coefficients are lumped in one, h_c. The two radiation heat transfer coefficients may also be lumped together in a single coefficient, h_r, and the above equation yields the following for the heat transfer expression from a FBR to an external surface:

$$\dot{Q} = (h_c + h_r)A(T_{fb} - T_w). \tag{3.10}$$

The definition of the radiation heat transfer coefficient, h_r, is straightforward and is derived directly from the equation for radiation heat transfer:

$$\dot{Q}_r = h_r A(T_{fb} - T_w) = \varepsilon\sigma A\left(T_s^4 - T_w^4\right) \Rightarrow h_r = \frac{\varepsilon\sigma\left(T_s^4 - T_w^4\right)}{(T_{fb} - T_w)}, \tag{3.11}$$

where σ is the Stefan–Boltzmann constant, which is equal to 5.67 W/m²K⁴, and ε is the emissivity of the system.

A simplification of Eq. (3.11) may be made if the temperature of the solids is approximately equal to the bulk temperature of the interior of the fluidized bed, $T_{fb} = T_s$, to yield the following expression:

$$h_r = \varepsilon\sigma(T_{fb} + T_w)\left(T_{fb}^2 + T_w^2\right). \tag{3.12}$$

For an enclosed surface or for a large flat plate, the emissivity of the system may be written in terms of the emissivity of the fluidized bed and the wall:

$$\frac{1}{\varepsilon} = 1 \bigg/ \left(\frac{1}{\varepsilon_{fb}} + \frac{1}{\varepsilon_w} - 1\right). \tag{3.13}$$

The emissivity of the entire fluidized bed is higher than the emissivity of the solids. An empirical relationship for the two is $\varepsilon_{fb} = \varepsilon_s^{0.48}$ (Borodulya and Kovenski 1983). The emissivity of the wall is usually assumed to be equal to the emissivity of the material comprising the wall. However, some authors suggest higher values for the wall emissivity because the collisions of particles with the wall surface cause the polishing of the surface.

Even though the last three equations may be combined and solved to yield the radiation heat transfer in a FBR, experimental data have shown that such calculations suffer from high uncertainty. The uncertainty is related to the fact that solid particles close to a heat transfer surface have temperatures that are between the surface temperature and the temperature in the bulk of the FBR. This implies that the actual temperature difference, which is relevant to the radiation heat transfer, is less than $(T_{fb} - T_w)$ and, hence, the solution of the last three equations overpredicts the actual heat transfer. In order to correct this, Schlunder et al. (1987) suggested that the arithmetic mean temperature between the bulk bed and the surface be used as the radiative temperature of the fluidized bed. On the other hand, Baskakov (1985) suggested that an empirical radiation heat transfer coefficient be used: $h_r = 7.3\sigma\varepsilon_{fb}T_w^3$. Another suggestion by Baskakov (1985) is to use an empirical effective emissivity for the FBR. All these empirical expressions depend on the internal geometry and scale of the FBR and are not applicable to all sizes and all configurations of FBR systems.

Regarding the convective heat transfer coefficient, h_c, this is not easily defined from first principles. It must be emphasized that h_c is significantly higher than the convective coefficient of the fluid alone, h_f. Even though the conduction between the solid particles and the external surface is negligible, the motion of the particles in the FBR and the vigorous agitation of the fluid cause significantly higher heat transfer rates and coefficients for the fluid–solids system, and therefore, $h_c \gg h_f$. It must be noted that the contribution of the solid particles on the heat transfer process in FBRs is the vigorous agitation of the fluid and the significant enhancement of the overall convective heat transfer coefficient, h_c, not the negligibly small amount of conduction through the surfaces during particle–wall collisions.

The range of operational velocities for FBRs is between the minimum fluidization velocity and the terminal velocity of the particles. When the fluid velocity is below the minimum fluidization velocity, the bed is stationary (it is a packed bed) and not fluidized. When the fluid velocity becomes more than the terminal velocity of the particles, we have pneumatic conveying and the particles would be carried out of the FBR. The convective heat transfer coefficient is a strong function of the superficial velocity in the FBR. When fluidization starts at the minimum fluidization velocity, U_f^{min}, there is typically bubbling, and the particles in the bed are still very close together. Particle movement and large-scale fluid agitation is limited because of the close proximity of the particles. At this stage, the heat transfer coefficient, h_c, is low.

As the superficial velocity increases, there is more space between the solid particles to move. The agitation in the fluid is enhanced. At this stage, the interparticle distance increases and h_c also increases following the velocity of the fluid.

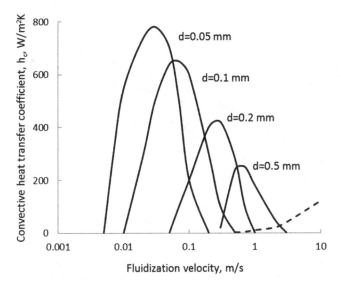

Fig. 3.3 Convective heat transfer coefficient, h_c, for different average sizes of particles and fluidization velocities

Experimental evidence has shown that the heat transfer coefficient increases as much as two orders of magnitude with the fluidization velocity and that the effect is higher with particles of smaller sizes. With any size of particles, there is a fluidization velocity; when the mixture becomes dilute, the average fluid volumetric fraction, $1 - \overline{\phi}$, becomes high, and the fluid space between the particles is too high to sustain the high convective heat transfer coefficients. Therefore, there is a fluidization velocity, above which the heat transfer coefficient starts decreasing and continues to decrease. When the fluidization velocity becomes close to the terminal velocity of the particles, the fluid would carry the particles outside the FBR. At this velocity the heat transfer coefficient, h_c, is close to the convective heat transfer coefficient in pneumatic conveying systems, which is significantly lower than the high coefficients observed in FBRs (Pfeffer et al. 1966; Michaelides 1986). Figure 3.3 shows this dramatic rise and drop of the fluid–solids convective coefficient, h_c, by several orders of magnitude to a channel inside the FBR for four particle sizes (Leckner 2006). The dashed line at the lower right represents the convective heat transfer of the fluid alone. The optimum performance of the FBR is at the point of maximum, h_c. It is apparent from this figure that the contribution of the particulates in the FBR increases the convective coefficient by almost two orders of magnitude. It is also apparent that a great deal of the fluidization advantage for heat transfer is lost for particles of sizes higher than 1 mm. The maxima in this figure have been correlated by the following expression (Varygin and Martyushin 1959):

$$h_c = \frac{0.43 k_f Ar^{0.2}}{\alpha} \quad \text{for } 30 < Ar < 10^5. \tag{3.14}$$

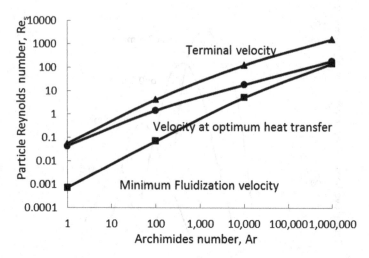

Fig. 3.4 The relationship between the Reynolds and Archimedes numbers for particles at minimum fluidization, onset of conveying, and optimum convective heat transfer

The fluidization velocity at the maxima has also been correlated by Gelperin and Einstein (1971) as follows:

$$Re_{\mathrm{opt}} = \frac{2\alpha U_{\mathrm{opt}}}{v_{\mathrm{f}}} = \frac{Ar}{18 + 5.22\sqrt{Ar}}. \tag{3.15}$$

The significant variability of the heat transfer coefficient in a FBR is one of the advantages for the use of FBRs as heat exchangers, even when there are no chemical reactions taking place. Figure 3.3 shows that an adjustment of the fluidization velocity between 0.01 and 0.1 m/s in a FBR with 0.05 mm particles would cause the modification of the heat transfer coefficient between 300 and 800 W/m^{2}K. This significant modification of the heat transfer coefficient represents a great advantage for a heat transfer medium.

Figure 3.4 depicts graphically the relationship between Ar and Re_{s} at the optimum convective coefficient for a FBR. The corresponding curves for the minimum fluidization and the terminal velocities of the particles are also shown in this figure. It is apparent that Re_{s} at the optimum heat transfer is close to the terminal velocity Reynolds number for small particles (low Ar) and gradually shifts to the minimum fluidization Reynolds number for larger particles.

It must be noted that both the radiative and the convective part of the rate of heat transfer in FBRs depend highly on the internal geometry and scale of the system. The empirical correlations and suggestions for the two heat transfer coefficients, h_{c} and h_{r}, are subjected to a significant degree of uncertainty. Given the high degree of empiricism in the calculation of the rate of heat transfer in FBRs, another approach is to disregard altogether the two components of heat transfer in Eq. (3.10) and express the overall rate of heat transfer in terms of an overall, average heat transfer coefficient, h. The latter combines the effects of both radiation and convection. Breitholz et al. (2001) determined that the density inside the FBR is the most important variable in the determination of the rate of heat transfer and suggested the following expression for the overall heat transfer coefficient:

$$h = 110\rho^{0.21} \text{ in W}/(\text{m}^2\text{K}), \tag{3.16}$$

where ρ is the spatially averaged density of the fluid–solids mixture in the FBR,

$$\rho = \rho_{\text{f}}\left(1 - \overline{\phi}\right) + \rho_{\text{s}}\overline{\phi} \quad \text{in kg}/\text{m}^3. \tag{3.17}$$

It is apparent from the above that the heat transfer in a FBR has a high degree of uncertainty and that it depends very much on the geometry and scale of the FBR. For this reason, the design of efficient FBR systems is more an art than a product of predictive science and relies on the experience of the designer, trial and error, and data from similar FBR models. The determination of the rate of heat transfer in actual industrial FBR designs is best accomplished by measurements of the heat transfer in scaled models and pilot plants.

3.4   Industrial Types of FBRs: Applications

FBRs have been used extensively in the chemical industry since the 1920s for coal gasification, combustion, pyrolysis, catalytic cracking, catalytic synthesis, hydrocarbon processes, metallurgical processes, roasting, calcination, biochemical processes, drug production, and purely physical processes, such as solids drying. In most cases, the fluidization systems are named according to the application they serve or the function they perform. For example, the Fluidized Catalytic Crackers (FCC), the Fluidized Bed Combustors (FBC), and the Fluidized Bed Gasifiers (FBG) are FBR systems used for the cracking of hydrocarbons, the combustion of coal, and the gasification of coal, respectively. Oftentimes in chemical engineering terminology, the part of the FBR that has uniform cross-sectional area, where most of the solid particles circulate and the majority of the chemical reactions occur, is called a *riser*. The riser is the main part of the FBR system.

The FBRs will work with any fluid, liquid, or gas. The majority of the FBR applications at present are with gases. In the sections that follow on industrial applications and systems, we will use the terminology of the gas reactors. It must be noted, however, that the same FBR systems and designs may be used with liquids.

3.4.1   Catalytic Cracking

The Fluidized Catalytic Crackers (FCC's) have become now an integral part of every crude oil refinery and provide millions of tons of additional transportation fuel daily. Their essential function is to split the heavy hydrocarbon molecules into molecules of light liquid products that have higher commercial value. An example of these reactions is the combination of the eicosane and lighter gaseous hydrocarbons to form three molecules of octane:

$$C_{20}H_{42} + 4CH_4 \rightarrow 3C_8H_{18} + H_2 \tag{3.18a}$$

or in the absence of the light hydrocarbon feed:

$$2C_{20}H_{42} \rightarrow 4C_8H_{18} + 5C + 3CH_4 \tag{3.18b}$$

This set of reactions, which are referred to as the *cracking process*, is favored at a range of medium temperatures close to 500°C and is facilitated by solid catalyst particles. Aluminum chloride was initially used as the catalyst in FCC followed by clay-based catalysts, which were used in the so-called *Houdry process*. The latter were substituted more recently with zeolite catalysts that provide maximum surface area for the hydrocarbon reactions. Regardless of the type of catalyst, the FCC operates at a temperature where the product feed (heavy and light hydrocarbons) is in the gaseous form and only the catalyst is solid. The fluidization process and the general agitation of the catalyst particles assist significantly in achieving a more homogeneous reacting mixture, which produces a consistent and uniform product.

In January 2010, FCCs had a total worldwide capacity of 13.2 million barrels of gasoline per day.[2] At a price of $3/gallon, this amounts to an economic contribution by the FCCs close to $1.7 billion/day. It becomes apparent that the optimum design of the FCCs, the flow patterns created, the local temperature, and local catalyst concentration are of paramount importance to the optimum operation of the FCCs and the economic value of its products.

One of the most important issues in the cracking processes is that the catalyst becomes poisoned after a certain time and needs to be "regenerated." The *regeneration* process occurs in a separate vessel and, typically, at a higher temperature. Therefore, a well-designed FCC allows for the poisoned catalyst to be continuously removed for regeneration and for the regenerated catalyst to continuously be fed back to the FCC. In well-designed FCCs, regeneration is accomplished by allowing the catalyst particles to be removed after a previously defined *residence time* in the FCC.

The regeneration process may be completed in another vessel, as shown in Fig. 3.5, where the regenerator and the cracking unit are stacked together. The cracker unit is at the top and the regeneration unit at the bottom of the figure.

---

[2] Worldwide Refinery Processing Review, 4th Quarter, 2009.

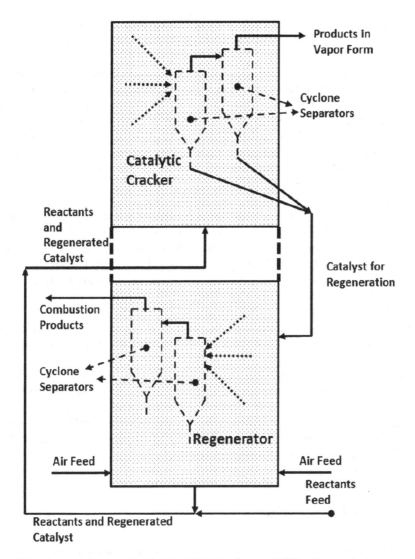

Fig. 3.5 Schematic diagram of a Fluidized Catalytic Cracker (FCC) with a catalyst regenerator unit

The reactants, which are often referred to as the *feed*, are vaporized and are fed at the bottom of the FCC unit to the cracker unit. As they rise, they also lift the regenerated solid catalyst particles and enter the cracker unit. The latter is at almost uniform temperature, close to 500°C, and is designed to allow the feed particles and the catalyst particles a predefined average *residence time*, during which they react and form the desired products. The gaseous products exit the catalytic cracker unit at the top. Cyclone separators and filters separate the catalyst particles from the product of the reactor. The catalyst particles are removed from the cracker unit by a pipeline, which connects to the side of the unit and is often called *the stripper*. The function of the stripper is to remove any products that are attached to the catalytic particles by a very slow-moving current of steam.

The catalyst particles fall by gravity in the stripper, where they are further compacted and fall to the vessel underneath, which is the regenerator unit. The temperature in the regenerator is maintained at a higher level—close to 600°C in clay-based catalysts and close to 700°C in zeolite catalysts—by burning the coke/carbon particles, which have been formed in the cracker unit and are carried by the catalyst particles. If more fuel is necessary for the regeneration process, CH_4 or C_2H_6 mixed with air for combustion is supplied to the regenerator unit. The regenerated catalyst is allowed to settle at the bottom of a feed pipe. A rotating gate valve facilitates the entrance of the catalyst to the feed line, which leads to the cracker unit. The excess air and combustion products exit at the top of the regeneration unit. A system of cyclone separators removes all the catalyst and coke particles before the flue gases exit this unit. As it may be seen in Fig. 3.5, the entire system receives the feed of heavy hydrocarbons and air (at the bottom right) and produces the flue gas from the combustion process (at the middle left) and the desired product of alkanes in a vapor form (at the top right). The products are primarily C_8H_{18}, C_7H_{16}, and C_9H_{20}, which are cooled to form the hydrocarbons in common gasoline. All intermediate processes, including the regeneration of the catalyst, are accomplished inside the FCC. This makes the system very convenient for all the complex processes involved in the cracking of hydrocarbons.

It must be noted that the system depicted in Fig. 3.5 is only one of the several FCC system designs and configurations that exist worldwide. While the configuration of the entire system may be different, the functions of the cracker and the regenerator units are the same in all systems. Solids fluidization, fluid agitation, and the homogenization that is accomplished in a FBR are essential characteristics of all the FCC designs that currently operate worldwide.

3.4.2 Catalytic Synthesis

The principal function of the catalytic synthesis FBR is to use the convenience and superior mixing characteristics of a fluidized bed in order to produce a uniform chemical product, whose molecules are more complex than the molecules of the reactants. The solid particles serve as catalysts and also as carriers of thermal energy. The continuous movement of the solid particles contributes to the homogenization of the temperature and species concentrations within the reactors. This facilitates the rate of the reaction and also ensures that the maximum yield is achieved for the desired reactions. In order to maximize the reaction yield, some catalytic synthesis reactors operate at high pressures. Very often the chemical reactions are exothermic or endothermic, which implies that significant quantities of heat must be exchanged between the gas–solids mixture and an exterior surface. The significantly higher heat transfer coefficients associated with the FBRs are of

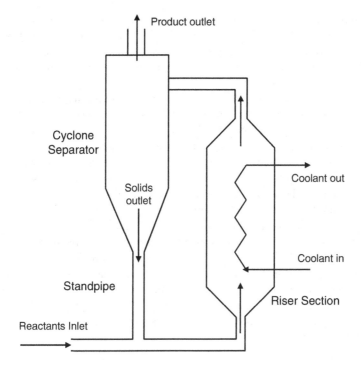

Product outlet

Cyclone
Separator

Solids
outlet

Coolant out

Coolant in

Standpipe

Riser Section

Reactants Inlet

Fig. 3.6   Schematic diagram of a recirculating reactor for catalytic synthesis

paramount importance in the exchange of heat and the maintenance of constant temperature in the reactor. Because of the vigorous agitation of the gas–solids mixture, a well-designed FBR transfers the maximum amount of heat for a given heat exchange area.

Catalytic synthesis reactors are used for the formation of a variety of chemical products. The production of synthetic transportation fuels, polymers, anhydrites, and acrylates are a few examples of these synthesis processes. The production of transportation fuel from CO and H_2 or CH_4 is a system of chemical reactions that are known by the generic name the *Fischer–Tropsch (F–T)* process. Solid catalyst particles that have been used are iron-based or silica–alumina particles, with the latter having several advantages because they fluidize easier. The F–T process is highly exothermic, which implies that a high rate of heat must be removed from the reactor.

In the 1950s, the South African Synthetic Oil Limited (Sasol) corporation developed a high-yield recirculating FBR for the F–T process, which became the prototype for the development of other, similar reactors. This type of reactors has a riser section, where most of the chemical reactions occur, and a downer, which returns the catalyst particles to be mixed with the reactants feed, as shown in Fig. 3.6. Because the reactions in the F–T process are highly exothermic, a heat exchanger system removes the heat at the middle section of the riser, where most of the reactions occur. This series of chemical reactors operate at intermediate pressures, between 10 and 17 bar. The reactor temperature is maintained at approximately 320°C. The volumetric fraction of solids, ϕ, in the riser section of these reactors is rather low, between 3% at the top and 10% at the bottom. This implies that the mixture is dilute and fluid turbulence plays an important role in the riser section, where most of the reactions occur.

The commercial success of the Sasol recirculating bed reactor and the significant technological expertise associated with this series of reactors has led many designers of chemical equipment to advocate recirculating reactors for all the F–T and most of the other catalytic synthesis processes. However, a moment's reflection will prove that, if the catalyst in the F–T process is not poisoned (and the majority of the catalysts in such processes are not poisoned), the solids do not need to be recirculated. Therefore, a well-designed non-recirculating reactor operating with the readily fluidized silica–alumina particles would have similar performance characteristics. In addition, the simpler non-recirculating bed reactor is easier to control and less expensive to design, construct, and operate. Given that most of the reactions in the F–T process as well as in most synthesis processes take place in the turbulent flow regime, and that a great deal of expertise has been acquired with non-recirculating FBR for chemical synthesis, a recirculating FBR is not necessary for the catalytic synthesis of materials. A simple FBR has been proven to be sufficient for such catalytic synthesis processes (Shingles and McDonald 1988).

3.4.3   Thermal Cracking and Coking

Thermal cracking involves the breakup of long hydrocarbon molecules for the production of shorter molecules of alkanes and alkenes, most common of which are ethylene (C_2H_4), propylene (C_3H_6), and butylene (C_4H_8). These chemicals are used as feed materials for the production of polymers that are commonly called *plastics*. The process is typically accomplished at very high temperatures (700–800°C) and does not involve a catalyst. Carbon is typically produced during this chemical process. A bed of inert solid particles keeps the temperature uniform by supplying the necessary heat for the endothermic reactions of thermal cracking. The addition of small amounts of catalysts serves to shift the chemical reactions to a desired product mix. For example, the addition of a catalyst and an increase of the temperature facilitates the ethylene formation in the reactions, and the product mix contains a higher percentage of this chemical. Thus, the addition of a suitable catalyst or a group of catalyst particles allows for significant flexibility in the product mix from the chemical reactor. This is a very useful tool for meeting changing industrial and market demands (Zhu and Cheng 2006).

Coking also involves the breakup of long molecules of hydrocarbons. The term is reserved for the breakup of the heavy residual byproducts of the refineries, such as bitumen from oil sands or residual material from the atmospheric distillation column of the refineries. Because the long hydrocarbon molecules invariably have more carbon atoms, the coking process produces free carbon in the form of coke.

A FBR is ideal for the coking process because the carbon particles that form the coke are carried in the reactor and facilitate mixing and higher product yield. In most of the cases, a fraction of the coke particles are burned inside the system to produce the heat needed for the endothermic reactions that take place. Steam is used in several coking units to separate the product. In some of these units, where there is excess carbon/coke, the steam is used to react with the coke particles and produce a synthetic gas, which is often called *coke gas*.

3.4.4 Fluidized Bed Combustors

Fluidized Bed Combustors (FBC) are increasingly being used in the power production industry, because they allow for a better controlled, more uniform combustion of coal, and require significantly less heat transfer area. Coal contains some solid particles that form the ash, but other inert solid particles, typically sands or clays, are added in the FBC. These particles provide sinks for the heat released from the combustion, maintain a uniform temperature, preheat the small coal particles when they enter the FBC, and increase the heat transfer coefficient to the water/steam that powers the turbine-generator system.

In addition, the mixing of the coal feed with limestone allows the in situ capture of several atmospheric pollutants, including sulfur dioxide (SO_2). The latter reacts with the limestone particles and forms the solid $CaSO_3$, which is discharged with the ash that is produced from the coal:

$$SO_2 + CaCO_3 \rightarrow CaSO_3 + CO_2 \qquad (3.19)$$

The introduction of limestone in modern FBCs as well as of sulfur scrubbers in older coal power plants has reduced significantly the amount of SO_2 released in the atmosphere. This practice has almost alleviated the problem of acid rain, which was a major environmental issue in the 1970s in North America and Europe (Michaelides 2012).

One of the main advantages of the FBC systems, in comparison to other coal burners and boilers, is that the interior temperatures are lower, typically in the range 750–900°C. In contrast, the boiler temperatures of representative coal power plants are close to 1,100°C and in several areas of the boiler exceed 1,200°C. The lower temperatures of the FBCs are translated in two significant advantages for the coal combustion process: The first is that alkali metals are not vaporized to exit with the flue gas, but remain as solids in the FBC and are removed with the ash. The second advantage is related to the nitrogen oxides, whose formation is facilitated at higher temperatures. These oxides are the entire series of nitrogen–oxygen compounds—NO, NO_2, N_2O_4, N_2O_3, etc.—which are usually denoted as NOX. Nitrogen oxides are precursors to ozone formation, are considered environmental pollutants, and are controlled by most environmental agencies, including the Environmental Protection Agency (EPA) in the USA.

The nitrogen oxides, NOX, are formed from the nitrogen compounds that naturally exist in the coal, such as ammonia, and not from the nitrogen supplied with the combustion air. The nitrogen compounds in the coal are typically released as volatile gases even at lower temperatures and are readily oxidized in common burners. During this oxidation process, nitrogen combines easily with oxygen to form NOX because it is released in the atomic form, N, not the molecular form, N_2. Herein lays the environmental advantage of the FBC systems: The separation of the air in two streams creates a chemically reducing zone at the bottom of the FBC, which does not favor the formation of NOX, even from atomic nitrogen. In addition, the partial combustion of the fuel in this part of the CFBC maintains lower temperatures. At these conditions of lower temperatures, the volatiles in the coal, including the volatile nitrogen compounds, are released and are partly oxidized. The prevailing lower temperatures and scarcity of oxygen do not favor the oxidation of nitrogen, which at this stage tends to form molecular nitrogen, N_2, rather than NOX. When the flue gases are lifted in the upper part of the riser section of the FBC, where the combustion is completed and the temperature rises, the nitrogen remains in the molecular form, N_2, and more NOX is not formed, despite the higher temperatures in this part of the reactor. Thus, the net effect of the separation of air in two streams is the complete combustion of coal and its volatiles. The combustion is achieved with significantly lesser production of NOX gases. Given the more stringent environmental emission standards globally, this characteristic makes all the FBC systems very promising for the next generation of coal burners.

While simple FBCs have performed very well in coal combustion, in the last thirty years, the Circulating Fluidized Bed Combustors (CFBC) have been introduced as a better alternative for coal combustion. Figure 3.7 depicts schematically the arrangement of a typical CFBC system. The CFBC is composed of a large diameter FBR, often called the riser, and a large cyclone separator next to it. The combustion air is fed primarily in two locations (a) below the distributor, where it causes the bubbling of the solids, and (b) midway through the riser section, where it facilitates the full suspension of the particles and brings the combustion process to completion. The solids fraction at the bottom of the riser is high. This fraction decreases gradually, approaching almost dilute flow at the higher parts of the riser. The finer solids are removed in the cyclone, and the flue gas is allowed to exit. Because the flue gas temperature is high, it passes through a heat exchanger, where it preheats the water for the power cycle. The separated solid materials are fed back to the riser by gravity, where any unburned fuel is burned and the ash is removed at a location close to the distributor plate. Additional heat exchangers—preheaters—may be installed at the inclined pipe of the cyclone to remove the heat from the returning particles and the ash.

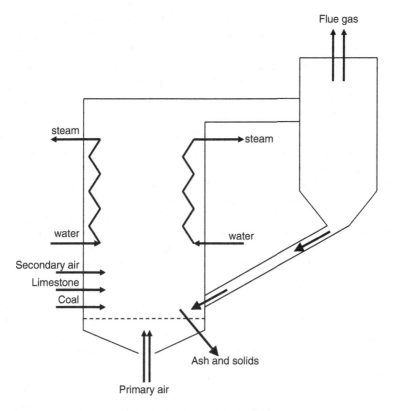

Fig. 3.7 Schematic diagram of a circulating fluidized bed combustor

3.4.5   Other Chemical Applications

It is apparent from the above that, because of their excellent heat and mass transfer characteristics, FBRs may be used in all chemical applications where products are formed in very large quantities. Among the other applications where types of FBRs have been used are the following:

Coal gasification: For this process, the coal is pulverized into smaller particles with sizes close to 1 cm and is fed into a FBR, which is also supplied by a mixture of air and steam. The coal reacts with the steam and is partly oxidized by the oxygen to produce CO and molecular hydrogen, H_2. High temperature is needed for these reactions to occur, and this is achieved by the partial oxidation of the carbon in the coal. The product, which contains a significant amount of hydrogen, is often used as feed for the production of other chemicals, e.g., ammonia, NH_3. Alternatively, the gaseous product may be burned in a conventional combustion chamber for the production of electric power. The gasification of coal before its combustion removes most of the environmental pollutants in solid form—sulfur, heavy metals, alkali metals, ash, etc.—and provides the burner with a cleaner fuel, which has significantly less environmental impact. In addition, the fuel, which is in gaseous form, may be easily transported to be used instead of natural gas, as *synfuel*, and also may be used in the combustion chambers of gas-turbine cycles operating with the Brayton cycle. One significant disadvantage to the gasification of coal, instead of direct combustion in a CFBC, is that fine coal particles are trapped both in the gas product and in the solid waste product (ash). This loss of carbon reduces the overall combustion efficiency of the gasifier-power plant system.

Solid waste incineration: Solid municipal waste has a heating value between 4,000 and 8,000 kJ/kg (1,715–3,430 Btu/lb) and is touted as an alternative fuel for the production of electric power. Industrial waste products typically have higher heating values. FBRs are ideal for the combustion/incineration of solid waste because they are designed to handle solids of several sizes and the combustion process is well controlled. One of the disadvantages of the currently used solid incinerators is that there is no adequate control for gaseous pollutants, such as heavy metal oxides, chlorine, and bromine compounds, and dioxins. When methods are developed for the elimination of these gaseous pollutants, FBRs will become ideal systems for power production from solid waste products.

Biomass combustion and pyrolysis: Biomass is another alternative fuel, which is produced from agricultural products or algae. Again, the controlled conditions in a FBR provide an ideal system for the combustion of biomass, either by itself or in a mixture with coal. The final product is process heat or electric power.

Mineral and metallurgical processes: FBRs are excellent candidates for chemical reactors that require good mixing and high, uniform temperatures. Roasting and calcination are two such processes. *Roasting* is a metallurgical process involving several gas–solid reactions at high temperatures. The purpose of roasting is to pretreat and partly purify a mineral before it is chemically treated to produce a chemical for the market. Examples of roasting are the treatment of arsenopyrite to obtain a porous solid, which is subsequently treated with cyanide to produce gold, and the oxidation of zinc sulfide to produce zinc oxide, which is subsequently used in electrolytic cells. *Calcination* is also a pretreatment process for the production of chemicals. During calcination, hydrates and hydroxides reject excess water from their molecules to produce refined chemicals for further processing or purification. For example, the calcination of aluminum hydroxide, $Al(OH)_3$, produces alumina, Al_2O_3:

$$2Al(OH)_3 \rightarrow Al_2O_3 + 3H_2O \qquad (3.20)$$

The latter is a dry powder, which is subsequently fed to electrolytic cells for the production of aluminum metal. Natural gas or coal may be mixed with the solid inside the FBR and be burned in order to supply the heat that is needed for the elevated temperatures in the FBRs and for all endothermic reactions that constitute the calcination and roasting processes.

Fig. 3.8 A solids dryer with
perforated baffles. The
cyclone at the top removes
any fine solids entrapped in
the air stream

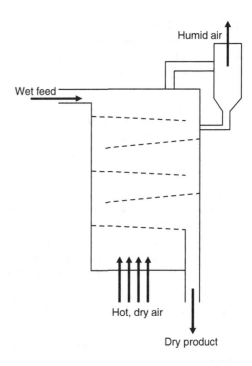

3.4.6   Nonchemical Applications

Drying of solid materials is one of the most common physical applications of FBRs
and is used extensively in the mineral, food, and pharmaceutical industries. The
advantages of using a FBR for the drying of solid particles vis-a-vis other drying
methods are:

1. Continuous feed of the solid particles and continuous removal of the dry product.
   This allows large-scale operations and systems.
2. Better control of the drying process.
3. Uniform temperature, which can be kept below the temperature that would
   damage the solids. This is particularly important for foodstuffs that may be
   spoiled at high temperatures.
4. The very large area of the suspended solid particles accelerates the drying
   process.
5. The agitation of the flow field by the particles establishes almost uniform and
   almost isothermal conditions in the dryer.

Figure 3.8 is a schematic diagram that depicts the operation of a dryer. Typically
the fluidizing air is heated by the combustion of gas before it enters the dryer. The
air outlet passes through a cyclone separator (and a filter in the pharmaceutical
production processes) to remove any fine particles that are carried by the air.
In order to control the quality of the product, oftentimes baffles are placed inside

the FBR. The baffles modify the velocity field and are designed to keep the particles in the drier for a uniform *residence time*. Such dryer designs ensure that the product dries without becoming overheated.

Oftentimes, the feed material has characteristics that complicate the drying process, e.g., it has a wide particle size distribution; is sticky or forms a paste; or has low cohesive strength. *Vibrating Fluidizing Bed Dryers* may be used in such cases. The combined fluidization and vibration allows the more gentle transportation of the materials that have low cohesive strength. For example, vibration is used extensively in dried dairy products and dried food processing.

Particle coating: The coating of particles is accomplished by, first, fluidization of the solid particles and, secondly, spraying the particles with drops of the coating material dissolved in a liquid solvent. When they come in contact with the solid particles, the drops form a uniform and thin layer around them because of the surface tension forces. This layer subsequently dries and forms the coating material. A suitable solvent for the coating is one that has high enough surface tension to form a thin, uniform layer around the particles and dries fast.

Powder coating: This process is used for the coating of larger metal objects with a layer of plastic material that protects them from erosion and corrosion or gives them increased mechanical strength. When a thermoplastic cover is used, the larger metal object is typically heated to a high temperature and dipped in a fluidized bed, where the thermoplastic particles have been fluidized. After a few seconds of this immersion, the surface of the hot metal is covered by the thermoplastic, and the entire piece is removed. Thermosetting powder coating materials are usually applied with a spray. Oftentimes an electrostatic spray method is used.

Heat exchange: Fluidized beds have much higher heat transfer coefficients, h_c, and higher volumetric heat capacities, ρc_p, than gases or liquids. These two properties make them ideal for heat exchange processes. For example, the shell side of a shell-and-tube heat exchanger may be a fluidized bed in a dense phase, which has excellent heat transfer characteristics as shown in Sect. 3.3. Fluidized beds may also be used as heat transfer materials from highly exothermic or endothermic reactions. In the latter case the fluidized bed, which is nonreactive, surrounds as a "blanket" the reaction vessel, absorbs all the heat that is generated, and, because of its high volumetric heat capacity, maintains almost constant temperature in the reaction vessel. An added advantage of a fluidized bed "blanket" is that its heat transfer coefficient may be adjusted at will by simply adjusting the volumetric flow rate of the fluid. This is demonstrated in Fig. 3.3, where it is observed that the variation of the fluidization velocity in the range 0.01–0.1 m/s (in the FBR with 0.05 mm particles) causes modifications of the heat transfer coefficient in the range 300–800 W/m²K. This significant adjustment of the heat transfer coefficient is a great advantage for any heat transfer medium. Essentially, the rate of heat transfer in and out of the FBR heat exchanger is controlled by the fluidization velocity, which is an easy parameter to control.

3.5 Computer Modeling: The MFIX Code

The processes of mass, momentum, and energy interactions in FBRs are very complex and include effects in the micro-, meso-, and macroscales. At the microscale, we have fluid–particle interactions, particle clustering, attrition, and particle segregation effects. At the mesoscale, we have the development of local fluid dynamics structures, such as jets and vortices, fluid bubbles in a solid matrix, erosion of solid surfaces, instabilities, and turbulence. All the processes of engineering interest, such as heat transfer to a working fluid or the production of chemicals, occur at the macroscale. Of course, the micro- and mesoscale phenomena are the causes of most of the effects at the macroscale. Therefore, the modeling of the engineering processes in FBRs, which are at the macroscale, must be linked to the phenomena at the other two scales, and this is done by using appropriate closure equations in the macroscopic models. In addition to the complex interactions, the geometry of all the commercially used FBRs is complex. The accurate modeling of these systems requires three-dimensional modeling.

At present, it is not feasible for a single numerical code to completely model the fluid–particle behavior in all the scales and to account for all the features of a FBR system. Several codes have been developed in the past two decades that model satisfactorily some of the most important macroscale effects in FBRs. Almost all of these codes are using the Eulerian, homogeneous model approach or the Eulerian, two-fluid approach as the sets of governing equations for the fluid–particle behavior (Sects. 2.4.1 and 2.4.2). A few of these codes are commercially available, such as the codes by FLUENT® and CD-Adapco®. Several of the industrial codes are proprietary and have been developed by corporations. However, most of the numerical codes have been developed in the academic research environment, are specialized and pertain to a specific research problem, a FBR type, geometry, or process.

It is important to know that all the available numerical codes are not comparable. The first task for the user of such a code is the full comprehension of the structure of the model used and of the inherent assumptions. Knowledge of all the governing and closure equations is of paramount importance. The users have to discern from the beginning what are the important features in their system that need to be modeled and to ensure that the code actually models accurately these features. For example, if the solids in the FBR have a wide distribution of sizes, a homogeneous model or a two-fluid model that only handles 3–4 sizes of particles is not the numerical model to be used. If particle aggregation and separation are important, the code must have a reliable module that accounts for particle interactions, will model these processes well, and will predict the correct drag coefficients and heat transfer coefficients for the aggregates.

All the reliable numerical codes for FBRs must be three-dimensional and must have the capability to allow for all the salient geometric features and equipment in the interior of the FBR, such as heat transfer channels, cyclone separators, and particle feeders. It is also important for the codes to model the entire FBR domain, including the inlet feeders and the outlets, not just the riser of the FBR under the assumption that the entrance and exit conditions are known or can be reasonably deduced. Actually, the accurate prescription of the inlet and outlet conditions is of paramount importance in the modeling of the FBRs. Because the governing equations in all the FBR models are parabolic, any uncertainties in the inlet and outlet conditions propagate in the interior of the riser and determine the solution of the model equations. Inaccurate inlet or outlet conditions will produce inaccurate or even meaningless results ("garbage in garbage out").

It is axiomatic that a numerical code is as good as the governing and closure equations it uses. Accurate closure equations provide reliable numerical results, and the opposite is true for inaccurate or uncertain closure equations. However, there are several processes and phenomena in the FBRs that are either not well under-stood or the closure equations, which describe them, are not accurate for all particle types or all chemical processes. An example of these phenomena is the heat transfer in the FBRs. Both the radiative and the convective heat transfer coefficients are not known to a high degree of accuracy, especially in dense fluid–particle systems. The mass transfer between the fluid and the particles also suffers from a high degree of uncertainty, especially where the density of particles is high. If chemical reactions are important, the numerical code must model accurately both the slow and the fast reactions. In most cases, the slow reactions are modeled more accurately. Fast reactions are significantly affected by the dynamics of the flow, which determine the distribution of the particles and the concentrations of the chemical species in the fluid. A species cannot react in a location of the FBR if it has been depleted in that location. Therefore, the kinetics and equilibrium of fast reactions must be coupled with the particulate flow and heat transfer processes.

Most of the effort in the fundamental modeling of the FBRs has been spent in the hydrodynamic interactions of the fluid and the particles. For that reason, the majority of the closure equations for the interaction of fluid and particles, e.g., the particle drag law, have been developed for isothermal flows. However, almost all the FBR applications pertain to chemical reactions and heat transfer, where the particles are at different temperatures from the carrier fluid. In general, particles and fluids interact differently in non-isothermal flows. Even for the simple case of the drag coefficient of a single particle in an infinite fluid, it has been shown (Feng and Michaelides 2008, 2009) that the fluid–particle drag is different when the particles are at different temperatures and natural convection develops around the hotter or colder particles. Absence of consideration of the chemical and thermal effects makes the results of the numerical code less accurate and sometimes meaningless. A desirable feature of a numerical code for FBRs would be to account for the fact that particles and fluid have different temperatures and that all the closure equations of the code reflect this.

Finally, the size of the computational grid must be considered. For Eulerian codes, the size of the grid must be large enough to contain a large number of particles and small enough to allow for good spatial resolution. In addition, too fine 3D grids require significantly higher computational resources that may not be readily available. For commercial large-scale units, the grid size may have to be of the order of 20–30 cm. Such a grid size will not offer good resolution and cannot account for local effects and phenomena. In addition, the modeler must check that the closure equations, which are used in the code, are applicable to larger scales and domains. Most of the empirical equations, especially the ones used for the heat and mass transfer processes, have been developed in studies pertaining to small, pilot-scale reactors or small computational domains. These empirical closure equations should be first validated with larger size reactors before their accuracy can be guaranteed.

Clearly, a great deal of fundamental modeling work needs to be done in order to ensure that we have accurate tools to model the complex processes in the FBRs. The following section will concentrate on the description of one of these tools, the MFIX code, which has been developed in the last decade as a result of a coordinated effort of many scientists and engineers in several universities and national laboratories.

3.5.1   The MFIX Numerical Code

Of the plethora of multiphase numerical codes that have evolved in the last twenty years, primarily in academic institutions and national laboratories, the MFIX (Multiphase Flow with Interphase eXchanges) code deserves special consideration because it is an open-source software and has been used and modified by several research groups. The development of the code started in 1996 in the National Energy Technology Laboratory (NETL) and has continued with contributions by several research laboratories, primarily in the USA. The code is free to use; it is of general purpose and models the momentum, energy, and mass interactions (including chemical reactions) of dilute and dense fluid–particle systems. It may be used simply in a single computer, or in a parallelized version, in several machines. In the last fifteen years, it has been used to describe several aspects of FBRs including bubbling, spouted, and circulating fluidized beds.

The MFIX code is based on a Eulerian, two-fluid approach. MFIX is written in FORTRAN, and among its capabilities are the following (Syamlal 1998):

1. It may handle multiple particle types and sizes.
2. It may be applied in three-dimensional Cartesian or cylindrical coordinate systems with uniform and nonuniform grids.
3. It accounts for the momentum and energy balances as well as for the gas and solids species balances.

The MFIX computations give time-dependent information on the following variables: pressure; velocity; volumetric fraction distributions for the fluid and solids; temperature; and chemical composition of the fluid and solid phases.

The numerical technique, which is used in this code, is the semi-implicit scheme that uses an automatic time-step adjustment. The specific scheme used in the current version of the code has been adapted from the method SIMPLE

(Semi-Implicit Method for Pressure Linked Equations) which is described by Patankar (1980). Two modifications of this technique were introduced in MFIX:

1. It uses a solids volume correction equation, instead of a solids pressure correction equation. This helps with the numerical convergence when the solids volume fraction is low. The correction also incorporates the effect of solids pressure, which appears to stabilize the calculations in densely packed regions with high solids volume fractions.
2. It uses automatic time-step adjustments to ensure the highest execution speed.

To improve the accuracy of the code, second-order accurate schemes for discretizing the convection terms are used in MFIX. A documentation and description of the elements of the code with a set of the basic governing and closure equations that are used may be found in Syamlal (1998) or in the several documents that appear in the MFIX Web site. A user's guide and several other documents that may help the user may also be found in the Web site of the NETL.

The most significant advantage of MFIX is that it is an open-source code and anyone may make modifications on the physical model that is used. The openness and free distribution of the software has created a relatively large community of users/researchers who have provided feedback and helped to improve the software (Pannala et al. 2010). Depending on the problems at hand, users may add terms to the governing equations, change the parameters of the closure equations, or adopt their own closure schemes. For example, users may supply their own constitutive relations for the stresses, may use any boundary conditions that appear reasonable, or choose one of several turbulence models for the calculations. This "experimentation" with the code has been often reported in the scientific literature, and the multiphase flow community was able to see the results, make comments on their accuracy, and adopt the best models and practices. One notable result of this widespread use and experimentation with this software is the combination of the MFIX code with other numerical methods, such as the DEM model, which pertain to the mesoscale processes and give more detailed descriptions of the flow field locally (Galvin et al. 2010).

It must be noted, however, that while the MFIX code may be considered as the most flexible and accurate of the available software at present, it still needs several improvements to accurately model the complex physical and chemical processes that occur in FBRs. The improvements needed are not in the numerical structure of the code, but in the choice of the closure equations. This is constrained by our current knowledge of the physical phenomena that take place in FBRs. When these complex phenomena and interactions are well understood, the accuracy of the modeling and simulations with multiphase systems will be significantly improved. The accurate determination of the complex processes in FBRs will be a milestone in the science of Multiphase Flow and Heat Transfer and a scientific triumph for the many researchers who have contributed to this field.

Bibliography

Aerov ME, Todes AM (1968) Hydraulic and thermal fundamentals of the operation of apparata with stationary fluidized particle beds. Chimia, Lenningrad (in Russian)

Baskakov AP (1985) Radiative heat transfer in fluidized beds. In: Davidson JF (ed) Fluidization. Academic, London

Borodulya VA, Kovenski VI (1983) Radiative heat transfer between a fluidized bed and a surface. Int J Heat Mass Transf 26:277–289

Breitholz C, Leckner B, Baskakov AP (2001) Wall heat transfer in CFB boilers. Powder Technol 120:14–48

Ding J, Gidaspaw D (1990) A bubbling fluidization model using kinetic theory of granular flow. AIChE J 36:523–538

Ergun S (1952) Fluid flow through packed columns. Chem Eng Prog 48:93–98

Feng ZG, Michaelides E (2008) Inclusion of heat transfer computations for particle laden flows. Phys Fluids 20:1–10

Feng ZG, Michaelides EE (2009) Heat transfer in particulate flows with direct numerical simulation (DNS). Int J Heat Mass Transf 52:777–786

Galvin J, Gargh, H, Lee T, Pannala S (2010) Coupled MFIX-DEM: verification and validation. In: NETL multiphase flow science workshop, 4–6 May 2010

Geldart D (1973) Types of gas fluidization. Powder Technol 7:285–293

Gelperin NI, Einstein VG (1971) Heat transfer to fluidized beds. In: Davidson JF (ed) Fluidization. Academic, London

Grace JR (1986) Contacting modes and behavior classification of gas-solid and other two-phase suspensions. Can J Chem Eng 64:353–366

Grace JR (2006) Hydrodynamics of fluidization. In: Crowe CT (ed) Multiphase flow handbook. CRC, Boca Raton

Leckner B (2006) Heat and mass transfer. In: Crowe CT (ed) Multiphase flow handbook. CRC, Boca Raton

Mathur KB, Gishler PE (1955) A technique for contacting gases with coarse solid particles. AIChE J 1:157–162

Michaelides EE (1986) Heat transfer in particulate flows. Int J Heat Mass Transf 29(2):265–273

Michaelides EE (2012) Alternative energy sources. Springer, Heidelberg

Pannala S, Syamlal M, O'Brien TJ (2010) Computational gas-solids flows and reacting systems. Engineering Science Reference, Hershey, PA

Patankar SV (1980) Numerical heat transfer and fluid flow. Hemisphere, New York

Pfeffer R, Rosetti S, Liclein S (1966) Analysis and correlation of heat transfer coefficient and friction factor data for dilute gas-solids suspensions. In: NASA report TN D-3603. NASA, Washington, DC

Schlunder E-U et al (eds) (1987) Heat transfer design handbook. VDI, Dusseldorf

Shingles T, McDonald AF (1988) Commercial experience with synthol CFB reactors. In: Basu P, Large JF (eds) Circulating fluidized bed technology II. Pergamon, Toronto

Syamlal M (1998) MFIX documentation—numerical technique. In: Report to U.S. DOE, number DOE/MC31346-5824, January 1998

Varygin MM, Martyushin IG (1959) Calculation of the heat transfer surface in fluidized bed equipment. Eng Chem 5:6–9 (in Russian)

Zhu J, Cheng Y (2006) Fluidized bed reactors and applications. In: Crowe CT (ed) Multiphase flow handbook. CRC, Boca Raton

Chapter 4
Heat Transfer with Nanofluids

Keywords Nanofluids • Continuum • Properties • Viscosity • Conductivity • Brownian motion • Double layer • Enhanced heat transfer

4.1 Introduction

Nanofluids are suspensions of nano-size particles (typically 5–1,000 nm) in fluids. Several research projects in the late 1990s and the first decade of the twenty-first century indicated that the addition of small amounts of nanoparticles in common cooling fluids increases significantly the effective conductivity of these suspensions. While experimentally determined conductivity enhancements were in the range 10–50%, some early experiments showed enhancement values higher than 100% (Choi et al. 2001). Experiments on the mass transfer coefficients with nanofluids reported more dramatic results with maximum mass transfer enhancements in the range of two to six times that of the base fluid (Kim et al. 2006; Olle et al. 2006; Komati and Suresh 2009). The significantly enhanced transport properties of the nanofluids have enormous implications in industrial processes, such as the cooling of very small electronic components, which will comprise the next generation of computer chips, absorption of gases by liquid carriers, increase of the rate of gas–liquid chemical reactions, electricity generation, cooling of IC engines, directed-energy weapons, boiling under microgravity conditions, nuclear reactor cooling, and biomedicine. Because of the enormous industrial and economic potential of nanofluids, a significant amount of research was conducted during the first decade of the twenty-first century on the thermal properties and applications of nanofluids, hundreds of journal articles were written, and several conferences were devoted to the subject.

The salient heat transfer characteristics of nanofluids will be described in this chapter, which starts with a fundamental description of continua, the definition of thermodynamic and transport properties of heterogeneous mixtures, and molecular

E.E. (Stathis) Michaelides, *Heat and Mass Transfer in Particulate Suspensions*, SpringerBriefs in Applied Sciences and Technology, DOI 10.1007/978-1-4614-5854-8_4, © Springer Science+Business Media New York 2013 considerations in nanofluids. Likely mechanisms for the enhancements of the transport properties—especially of viscosity and thermal conductivity—are also described. Results on the thermodynamic and transport properties of the nanofluids are given, and the methods of measurements as well as the underlying enhancement mechanisms are critically described.

4.2 Continuum and Molecular Considerations

The concept of a continuum is a fundamental concept, which is central to the description of most engineering systems. By using the concept of a continuum, one may apply the principles and methodology of calculus to materials composed of discrete atoms and molecules. The concept of a continuum is based on several implicit assumptions pertaining to the definition of the local thermodynamic and transport properties, and the practical meaning of the mathematical operations of differentiation and integration, which are part of the constitutive and governing equations of all materials, including nanofluids. For this reason, a practical and simple exposition of the continuum concept is given here with the property density as an example (Michaelides 2006). The mass of a material of nonuniform density enclosed in a volume, V, is defined by the equation

$$m = \int_V \rho(x, y, z) \mathrm{d}V, \qquad (4.1)$$

where $\rho(x,y,z)$ denotes the density function of the fluid, a quantity that is normally nonuniform. This operation is based on the implicit assumptions:

(a) The density function of the material exists.
(b) The density function is well defined at every point of the continuum, which occupies the volume V.

It must be recalled that the local density function is defined mathematically by the limit operation:

$$\rho = \lim_{\Delta V \to 0} \frac{\Delta m}{\Delta V}, \qquad (4.2)$$

where Δm is the mass contained within a volume ΔV of space and the limit operation is defined as the volume, ΔV, becomes "arbitrarily small." Given the atomic structure of matter, one is faced with the paradox that, when the volume ΔV is reduced to a geometric point in space by becoming "arbitrarily small," there is a very low probability that an atom or part of an atom exists in this volume. Hence, the local density of the material at a point, $\rho(x,y,z)$, is most likely zero at the point (x,y,z). If the volume is sufficiently small and part of an atom actually existed in the

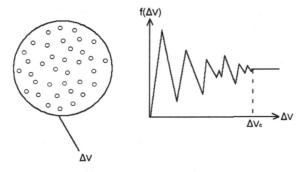

Fig. 4.1 The function $\Delta m/\Delta V$ in terms of the parameter ΔV

volume ΔV, then the density, as defined by Eq. (4.2) would be very large. If one adopted this operational definition for the material density, by considering "arbitrarily small" volumes ΔV, the density function would be highly nonuniform and the numerical values for density obtained using Eq. (4.2) would be meaningless. Because all molecules and atoms undergo some type of motion—vibrations about a fixed position in solids, almost random motion in fluids—the nonuniform density function would also exhibit a very fast variability with respect to time.

A moment's reflection proves that under these circumstances, when the volume ΔV is required to be "arbitrarily small" and its dimensions become of the order of magnitude of the atomic structure, it would be impossible to have an operational definition of the property density or for that matter of any other material property. However, if the volume, ΔV, may be defined to be large enough to contain a sufficiently large number of molecules, e.g., a few thousand, the fact that a few molecules move in and out of this volume has little effect on the mass Δm inside the volume, and the quantity $\Delta m/\Delta V$ converges to a limit, which may be defined as the operational density of the material.

Under this operational definition it is possible to assign a value for the density function $\rho(x,y,z)$ to every point in the material: The density is equal to the limiting value of the function $\Delta m/\Delta V$ and is reached when the volume ΔV is sufficiently small from the macroscopic point of view but still large enough compared to the molecular dimensions. When the mathematical function, $\rho(x,y,z)$, is defined at every point within the larger volume, V, the integration denoted by Eq. (4.1) may be performed and the mass of the material may be determined. It must be emphasized that, for the integration denoted by Eq. (4.1) to be performed, the density function only needs to be properly defined, and it does not need to be uniform, continuous, or differentiable.

Figure 4.1 is a schematic diagram that demonstrates this definition of material density. The left part shows the molecules of the material inside the volume ΔV, and the right part depicts the function $f(\Delta V) = \Delta m/\Delta V$. It is apparent that ΔV_c is the lowest value the volume ΔV may attain for the operational definition of the density function to be meaningful as a property of the continuum.

In a similar way, functions for the other material properties, such as the specific enthalpy $h(x,y,z)$ or the specific total energy $e(x,y,z)$, may be appropriately defined at every point of the material. All the operations that are defined in calculus may be performed with these properties. This implicit continuum assumption underlies all definitions, mathematical operations, and equations of continuum theory, which have become the foundation of science and engineering.

It must be noted that the validity of the continuum assumption does not stem from a physical principle or a mathematical proof but is inferred from the fact that the resulting "continuum description" of the materials does not conflict with any empirical observations and, actually, is supported by all the available empirical data for physical systems containing a large number of molecules.

In the cases where the continuum description does not apply, e.g., in rarefied gases and molecular films, one needs to seek a different way for the meaningful definition of the material properties at the molecular level. The Knudsen number, Kn, is a dimensionless quantity that gives an indication if a material may be treated as a continuum. In the case of nanofluids, Kn is defined as the ratio of the molecular free path of the fluid, L_{mol}, and the apparent diameter of the particles in the nanofluid:

$$Kn = \frac{L_{mol}}{2\alpha}. \tag{4.3}$$

If $Kn \ll 1$ the system may be described as a continuum. Given that the order of the molecular free path of most common liquids is less than 1 nm (10^{-9} m) and the vast majority of applications of nanofluids involves particles with sizes higher than 20 nm, the relevant Knudsen numbers are lower than 0.05 and typical liquid nanofluids may be treated as continua. In the cases where the base fluid is a gas, the mean free path of the gas is of the order of 10 nm. Particles must be larger than 100 nm for nanofluids with a gas base to be considered as continua.

The above considerations have significant ramifications when the solids may be considered as a second continuum as it happens in the case of the "two-fluid model" that is often used in numerical calculations (Sect. 2.4.2). For meaningful numerical computations to be performed with this numerical approach, the pertinent volumes, or the numerical grid, must be large enough for a large number of elements of the second phase—particles, drops, or bubbles—to be present.

4.3   Characteristics of Nanofluids

Particulate systems have been considered as heat and mass transfer media since the 1950s (Pfeffer et al. 1966) for the removal of heat from nuclear reactors. However, fluids with larger particles, of sizes of mm and above, are not suitable heat transfer media because large particles cause erosion and deposition on the walls. Nanofluids are composed of much smaller particles (less than 1 μm). These particles do not have significant sedimentation velocities, do not cause significant erosion to the equipment, and do not readily deposit at the wall boundaries.

The improved heat and mass transfer properties of the nanofluids are due to two factors (a) the significantly enhanced thermal conductivity of the fluid by the addition of nanoparticles and (b) the motion of the nanoparticles inside the fluid, which causes local agitation.[1] In addition to the advection of particles, the following characteristics that are unique to nanofluids affect their motion and heat transfer.

4.3.1 Surface-to-Volume Ratio

The surface-to-volume ratio is a geometric ratio that has important implications in the properties of fluid–particle mixtures. Let us consider a heterogeneous mixture of a fluid and a solid, both contained in a volume V. The fluid occupies a volume V_f and the solid occupies a volume V_s. Let us also assume that the solid is composed of N uniform spheres of radii α, and, hence, $V_s = 4N\pi\alpha^3/3$. The area of the solids that is exposed to the fluid is $A_s = 4N\pi\alpha^2$ and the surface-to-volume ratio is $3/\alpha$. Apparently, when the volume of the solid particles remains the same and the number of the particles, N, increases by making the particles smaller in size, the surface-to-volume ratio increases and is inversely proportional to the size of the particles. A consequence of this is that in nanofluids the total interfacial area of the particles is significantly higher than in other particulate systems with the same solids volumetric ratio. Therefore, all processes that depend on the interfacial area of the particles would be affected significantly. For example, chemical reactions, reaction catalysis, and absorption are surface processes that are proportional to the interfacial area and are expected to proceed faster in the presence of nanoparticles. Heat conduction is also a process that depends on the surface of the particles and is affected by the size of the particles.

4.3.2 Brownian Motion

This motion was first observed in a microscope by Robert Brown in 1837 and was described analytically by Albert Einstein in 1905. The Brownian motion is the aggregate result of all the impacts of the fluid molecules on the surface of the particles. The fluid molecules have significantly high velocities, of the order of 1,000 m/s, which depend on the temperature of the fluid. Actually the molecular velocities define the fluid temperature of a homogeneous material through the expression (Tien and Lienhard 1979)

---

[1] In addition to the motion of the particle, this agitation includes the movement of the fluid that is carried by the particle as virtual mass and also the movement of the surrounding fluid that "rushes in" to fill the volume of the moving particle.

$$T = \frac{m\overline{C^2}}{3k_B}, \tag{4.4}$$

where m is the mass of a molecule, C is the magnitude of the velocity of the molecules, and k_B is the Boltzmann constant, $k_B = 1.38 \times 10^{-23}$ J/K. A consequence of the last equation is that molecules of continua at higher temperatures have higher velocities and *vice versa*.

Molecular collisions with particles are random and take place at the molecular timescales, which are of the order of femtoseconds (10^{-15} s) and much shorter than the timescales of the particles. Because the mass of the individual molecules is very small in comparison to the mass of the nanoparticles, the impacts of the individual molecular collisions on the particles are of very small magnitude. However, because the number of impacts per unit time is very large, the aggregate effect of molecular impacts is noticeable on the movement of particles with sizes less than 10 μm and is considered negligible for larger particles. Brownian motion is particularly significant in nanofluids, where the transport properties are modified by this random movement of the constituent particles with respect to the bulk of the fluid.

Let us consider the motion of a small particle under the influence of the random molecular impacts in a system where all the other surface and body forces vanish. During time periods of the order of the momentum timescale of the particle, τ_M, the effects of the Brownian motion may be expressed as the action of a random force, $\vec{F}_{Br}$, which acts on the particle continuously. This force is counteracted by a Stokesian-type drag. While the individual molecular impacts are not in the scale of the continuum, one may consider the aggregate of the molecular effects on the Stokesian drag of the particle, which is at the continuum scale, to be expressed by a drag multiplier, f_{Kn}. The latter is a function of the Knudsen number (Michaelides 2006). Hence, the equation of motion of the particle may be written as follows:

$$m_s \frac{d\vec{v}}{dt} = 6\pi a\mu f_{Kn}(\vec{u} - \vec{v}) + \vec{F}_{Br}. \tag{4.5}$$

When the fluid velocity does not change with time (steady motion), the last expression yields the following equation of motion of the particle:

$$\frac{d}{dt}\left(\frac{d\vec{x}}{dt}\right) = -\frac{9\mu f_{Kn}}{2a^2 \rho_s}\frac{d\vec{x}}{dt} + \frac{3}{4a^3 \rho_s}\vec{F}_{Br}, \tag{4.6}$$

where ρ_s is the density of the solid material that comprises the particle. In order to quantify the effects of the Brownian motion on the transport properties of the nanofluid, one must quantify the random force, $\vec{F}_{Br}$, on the velocity and position of the nanoparticles. One way to accomplish this is to multiply the terms of Eq. (4.6) by $\vec{x}$ (scalar product) and take a time average of all the possible motions of the particle or, equivalently, the ensemble average of the motions of a large number of particles (Russel et al. 1989). If the ensemble average is denoted by the angular brackets, $<>$, the resulting averaged equation of motion may be written as follows:

$$\frac{d}{dt}\left\langle \vec{x}\cdot\frac{d\vec{x}}{dt}\right\rangle - \left\langle \left(\frac{d\vec{x}}{dt}\cdot\frac{d\vec{x}}{dt}\right)\right\rangle = -\frac{9\mu f_{Kn}}{2\alpha^2\rho_s}\left\langle \vec{x}\cdot\frac{d\vec{x}}{dt}\right\rangle + \frac{3}{4\alpha^3\rho_s}\langle \vec{x}\cdot\vec{F}_{Br}\rangle. \qquad (4.7)$$

The motion of the particles in the fluid does not have a preferred direction, and hence, the ensemble average of the random Brownian force vanishes. Also, the molecular impacts have timescales that are of much shorter duration than τ_M. During time periods of the order of the timescale of the particles, the second term in the LHS is equal to the equilibrium velocity fluctuations of the particles. This is equally partitioned among the three spatial directions. The value of this ensemble average is $3k_BT/2$. Hence, the expression for the ensemble-average displacement becomes (Russel et al. 1989)

$$\frac{d}{dt}\left\langle \vec{x}\cdot\frac{d\vec{x}}{dt}\right\rangle = \frac{9k_BT}{8\alpha^3\rho_s} - \frac{9\mu f_{Kn}}{2\alpha^2\rho_s}\left\langle \vec{x}\cdot\frac{d\vec{x}}{dt}\right\rangle. \qquad (4.8)$$

The last expression may be integrated twice with the initial conditions that the particle position and velocity are equal to zero to yield the following expression for the ensemble-averaged dispersion of the particle:

$$\langle \vec{x}\cdot\vec{x}\rangle = \frac{2k_BT}{6\pi\alpha\mu f_{Kn}}t. \qquad (4.9)$$

One may note that the particle dispersion, due to the Brownian motion, is independent of the density and the other characteristics of the particle and only depends on the size of the particle. For large-enough particles in a gravitational field, this dispersion is negligible. The dispersion coefficient, D_0, is equal to half the derivative of the Brownian dispersion. When the Knudsen number effects are negligible and only the Stokesian drag is considered ($f_{Kn} = 1$), one may obtain from Eq. (4.9) the so-called *Stokes–Einstein diffusivity* of an isolated spherical particle:

$$D_0 = \frac{d}{2dt}\langle \vec{x}\cdot\vec{x}\rangle = \frac{k_BT}{6\pi\alpha\mu}. \qquad (4.10)$$

Sometimes, e.g., for Monte Carlo simulations, it is desirable to perform a Lagrangian simulation for the motion of an ensemble of particles. A computational time interval Δt is chosen a priori for the numerical integration of the particle motion during the simulations. Δt is typically much higher than the characteristic time of molecular collisions with the particle, and hence, the time-averaged dispersion of the ensemble of particles is expected to be equal to the value predicted by Eq. (4.10). A simple way to incorporate the effects of the Brownian motion on the Lagrangian trajectories of the ensemble is to include in the computations a random force, $\vec{F}_{Br}$, whose sole effect on the ensemble of particles is to cause a dispersion equal to that predicted by Eq. (4.10) during a time interval Δt. This random force is the same as the force $\vec{F}_{Br}$ which was introduced in Eq. (4.5). For this force to cause the dispersion defined by Eq. (4.10), the force must be equal to the following expression:

$$\vec{F}_{Br} = \frac{4}{3}\alpha^3 \rho_s \vec{R} \sqrt{\frac{2k_B T}{6\pi\alpha\mu f_{Kn}(\Delta t)^3}}, \tag{4.11}$$

where $\vec{R}$ is a random vector, whose components are Gaussian random numbers with zero mean and unit variance. For completeness, the Knudsen effects on the particle drag were reintroduced in the last equation via the correction factor f_{Kn}.

4.3.3   Thermophoresis

Thermophoresis is an interesting consequence of the Brownian motion of the particles. As it is apparent from the last three equations, the particle dispersion is higher and the Brownian force is stronger within regions of higher fluid temperatures. When there is a temperature gradient in the particulate system, small particles tend to disperse faster in hotter regions and to disperse slower in colder regions. The aggregate result of this differential dispersion is the migration of particles from the hotter to the colder regions of the system and the relative accumulation of particles in the colder regions of the system. This is depicted schematically in Fig. 4.2, which shows the effects of the magnitude of the molecular collisions on small spherical particles. In steady-state systems, particle collisions in the colder regions, where the concentrations are higher, partly counteract this accumulation, and a dynamic equilibrium for the particle concentration is established, with lower concentrations in the hotter regions and higher concentrations in the colder regions.

The effect of thermophoresis on small particles is often expressed in terms of a thermophoretic velocity, v_{tp}, or of a thermophoretic force, F_{tp}. The two act in the direction opposite to the temperature gradient and are defined as follows:

$$v_{tp} = -K_{tp}\frac{\mu f_{Kn}}{\rho_f}\frac{\nabla T}{T_\infty} \quad \text{and} \quad F_{tp} = -6\pi\mu^2 f_{Kn}\alpha K_{tp}\frac{\nabla T}{\rho_f T_\infty}. \tag{4.12}$$

The function K_{tp} depends on the Knudsen number and the properties of the fluid and the solid particles. Michaelides (2006) gives several correlations of the function $K_{tp}(Kn)$ that apply to the different flow regimes of rarefied flows. Among the flow regimes, of interest in the case of nanofluids is the *continuum flow*, where both fluid and particles are treated as continua and the particles exert an influence on the fluid velocity field. Under these conditions, velocity and temperature

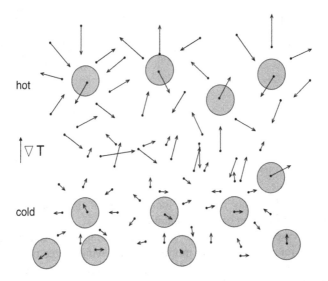

Fig. 4.2 Thermophoretic motion of spherical particles. The Brownian motion brings more particles to the colder region of the system

discontinuities (slip) are likely to manifest on the interface between particles and fluid. Thus, K_{tp} depends on the type of motion as well as on the discontinuities at the interface. By considering the velocity slip as defined by Basset (1888) as well as a temperature slip, Brock (1962) derived the following expression for K_{tp}, in terms of the Knudsen number:

$$K_{tp} = \frac{2C_s(k_f + 2k_sKn)}{(1 + 6C_mKn)(2k_f + k_s + 4k_sC_tKn)}. \tag{4.13}$$

The parameters, C_s, C_m, and C_t, are determined empirically from the flow field around the particles and the discontinuities on the fluid–particle interface. Talbot et al. (1980) used the velocity slip expression that was recommended by Millikan (1923) and derived empirical functions for these parameters in terms of "accommodation coefficients." Experimental data and engineering practice have shown that reasonable results may be obtained by treating these parameters as constants with the values $C_s = 1.17$, $C_m = 1.14$, and $C_t = 2.18$ when $Kn < 0.1$.

Although the thermophoretic force is a weak force and has vanishing effect on particles with sizes larger than 10 μm, it may become a dominant force in the case of very small nanoparticles where the gravity force is very weak. A glance at Eq. (4.12) proves that the ratio of the thermophoretic to the gravity force varies as α^{-2} and that the thermophoretic force would be dominant on nanoparticles in a gravitational field. Because of this, thermophoresis, rather than sedimentation, has often been used for the collection of nanoparticles on surfaces and especially of nanoparticles that are formed in vapor or gaseous streams in the process that is called *vapor deposition*. Since the thermophoretic force is a very weak force,

Fig. 4.3 A charged sphere
in an electrolytic solution—
electrophoretic velocity

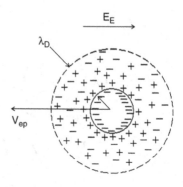

this collection method may take very long periods of time and requires precise
equipment and suppression of flow instabilities.

It must be emphasized that the thermophoretic force is a consequence of
molecular collisions and Brownian motion. The Brownian force of Eq. (4.11) and
the thermophoretic force of Eq. (4.12) have the same cause and they are not
independent forces to be modeled and accounted for separately. For this reason it
is erroneous to include both of them in an equation of motion of the particle.

4.3.4 Electrical Double Layer, Zeta Potential, and Electrophoresis

Most nanofluids comprise electrically charged particles and base fluids composed
of polarized molecules or electrolyte solutions. The position of the electric charge,
the electric forces on the particles, and the motion of the electric charge play an
important role in the structure of the particles and the transport properties of the
nanofluid. The electric force is a surface force, scales as α^2, and has a greater
influence on the behavior of fine particles than the body forces, such as gravity,
which scales as α^3. For this reason, the electric forces and their effect on particles
have been used extensively in the characterization, separation, or fractionation—
separation in fractions of different sizes—of polydisperse systems; the study of
surface properties; and the preparation of flocs and aggregates with desired composition and properties. Electric attraction is also used in the stabilization of colloidal
systems and the formation of gels.

Consider the negatively charged sphere in an electrolytic solution of concentration c_E shown in Fig. 4.3. An electric field of intensity E_E is imposed on the
solution. The sphere starts moving in the direction opposite to the field. Positive
ions from the electrolyte are attracted by the negative charges on the surface of the
sphere, aggregate at this surface, and form a layer around it. This layer is called
the Debye sheath, or more often the *double layer*. As the sphere moves in the
viscous electrolyte solution, it carries the double layer with it. This adds to the
inertia of the sphere. The extent of the double layer is characterized by the Debye
length, λ_D. The latter is a measure of the radial distance where the electric potential,

which is due to the presence of the sphere and the associated double layer, reduces by a factor $e^{-1} = 0.368$:

$$\lambda_D = \sqrt{\frac{\varepsilon\varepsilon_0 k_B T}{2{,}000\ e^2 z_E^2 C N_{av}}}, \tag{4.14}$$

where ε is the dielectric constant of the solution; ε_0 is the electric permittivity of vacuum, 8.85×10^{-12} F/m; e is the electric charge of an electron, $e = 1.6 \times 10^{-19}$ C; z_E is the charge number (valence) of the ions in the electrolytic solution; N_{av} is the Avogadro number 6.023×10^{-23} ions/mole; and C is the concentration of the solution in mol/L. For an aqueous solution at $25°C$, the last equation yields the Debye length in m:

$$\lambda_D = \frac{3.041 \times 10^{-10}}{z_E \sqrt{C}}. \tag{4.15}$$

It is apparent that the Debye length in aqueous solutions is on the order of a few nanometers, which is significantly smaller than the size of most nanoparticles that are used in typical nanofluids. Therefore, the effects of the double layer are confined within a very short distance from the surface of the particle. The mass of the fluid in this short distance is by far smaller than the mass of the particles.

One way to model the electric charge around the sphere according to the schematic diagram of Fig. 4.3 is to consider two concentric spheres with opposite charges, q_s and $-q_s$ at two radii α and $\alpha + \lambda_D$ from the center of the sphere. The potential created by these spheres is called the *zeta potential* and is given as

$$\zeta = \frac{q_s}{4\pi\varepsilon\varepsilon_0\alpha} - \frac{q_s}{4\pi\varepsilon\varepsilon_0(\alpha + \lambda_D)}. \tag{4.16}$$

In the limit $\alpha \gg \lambda_D$, which is often called the small Debye length limit, the relationship between the surface charge, q_s, and the zeta potential is

$$q_s = \frac{\varepsilon\varepsilon_0\zeta}{\lambda_D}. \tag{4.17}$$

The presence of the external electric field induces a motion on the electrically charged particles, which is called *electrophoresis*. Electrophoresis is resisted by the hydrodynamic drag on the particles. As a result, the particles attain steady motion with a constant velocity, the *electrophoretic velocity*, v_{ep}. Since nanoparticles are very small, they exhibit Stokesian drag. If the net charge around the nanoparticles is denoted by q_E, the balance of the electrostatic and viscous forces yields the following expression for the electrophoretic velocity:

$$q_E E_E = 6\pi\mu\alpha v_{ep}. \tag{4.18}$$

The net charge q_E is approximately equal to $4\pi\varepsilon\varepsilon_0\zeta$. Then from the last equation, an expression for the electrophoretic velocity may be derived in the limit $\lambda_D \gg \alpha$, which is (Probstein 1994)

$$v_{ep} = \frac{2\zeta\varepsilon\varepsilon_0 E_E}{3\mu}, \quad \lambda_D \gg \alpha. \tag{4.19}$$

In the opposite limit $\alpha \gg \lambda_D$, which is the case of aqueous media, one must account in detail for the influence of the fluid double layer on the motion of the sphere, the effect of neighboring charged spheres, and the motion of the counterions in the fluid. One may derive again an approximate expression for the electrophoretic velocity (Probstein 1994):

$$v_{ep} = \frac{\zeta\varepsilon\varepsilon_0 E_E}{\mu}, \quad \lambda_D \ll \alpha. \tag{4.20}$$

The last equation is sometimes referred to as the *Helmholtz–Smoluchowski equation*. Equations (4.19) and (4.20) are used to determine the value of the zeta potential of particles, by observing the motion of particles under a microscope in a fluid of known properties, usually deionized water.

It must be noted that when $\alpha \gg \lambda_D$ and the double layer is very thin, the electrostatic forces between the particles and the double layer are strong, and the nanoparticles carry with them the part of the fluid that is contained in the double layer with the constant velocity v_{ep}. In this case, the thin fluid double layer moves as a "plug" within the base fluid, and there is a velocity slip at the interface of the double layer and the base fluid. This has led some researchers to speculate that the fluid "solidifies" on the surface of the particle. However, thermodynamic equilibrium considerations do not support a phase transition in this double layer. The observed behavior of the fluid in the double layer to move as a plug is solely due to the strong electrostatic forces and not to a phase transition.

4.3.5 Aggregation and Separation of Particles

The combination of the electric charges on the surface of the particles and the dielectric properties of the base fluid and the particles has very important consequences on the aggregation of particles to form larger groups. The relative position of the particles in a fluid changes constantly because of:

(a) Bulk motion of the fluid
(b) Lift force in regions of high strain
(c) Electric forces between particles
(d) Hydrodynamic forces between pairs of particles or between particles and walls
(e) Brownian motion

Fig. 4.4 Interparticle
potential

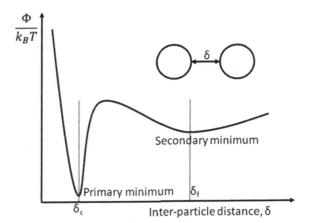

As a result, particles come to close proximity and form bonds that keep them in groups of aggregates, which may trap the interstitial fluid. Depending on the nature of the particles and their electrostatic properties, some nanofluids may appear as solutions and others as solid-like gels. The formation of aggregates of nanoparticles and their flow behavior influence significantly the structure and the transport properties of nanofluids.

The aggregation process of the particles is a very complex process and depends on several variables of the base fluid and the particles. One way to model the aggregation process is to combine all the forces acting on pairs of particles in a single potential function, Φ, which represents the net potential of all the attractive and repulsive forces acting on a pair of particles. Figure 4.4 shows a typical graph of the dimensionless form of this potential function, Φ/k_BT, against the interparticle distance, δ. It is observed that, as the interparticle distance increases, the potential energy function exhibits a high-energy primary minimum followed by a lower-energy local maximum and a low-energy secondary minimum. The two interparticle distances corresponding to the minima, δ_c and δ_f, define two stable configurations of the aggregates. Simple energetic considerations prove that when the interparticle distance corresponds to the primary minimum, δ_c, the bonds formed between the particles are very strong. This process is called *coagulation*.

The bonds corresponding to the secondary minimum, δ_f, are weaker bonds and the aggregation process at this point is known as *flocculation*. It is also noted in the figure that the local maximum or energetic "hump" that the particles have to overcome in order to move from δ_f to shorter interparticle distances and coagulate is fairly low. Particles may overcome and "jump over" this energetic hump to enter the coagulation process solely with the thermal fluctuations of their energy, which are characteristics of the Brownian motion. The energy deficit in "going down the hump" is transformed to the strong bonds that characterize the coagulants. The hump in the opposite direction, from δ_c toward increasing interparticle distances, is significantly higher, and thermal energy alone is not sufficient to overcome this barrier to break up the coagulant bonds. For this reason, coagulants are very stable, do not break up easily, and behave as larger solid particles in a flow field.

As the shallow secondary energy minimum implies, the interparticle bonds in the flocs are significantly weaker. Flocs may break easily with the weaker hydrodynamic forces, which include shear forces, turbulence, interactions and collisions with other particles, and collisions with the flow boundaries. Thus, the formation and breakup of flocs are affected by the flow boundaries and the type of flow. Experimental results show that the formation of flocs determines the structure and transport properties of nanofluids, such as viscosity, thermal conductivity, and mass diffusivity. For this reason, the flocculation process and the dynamics of the interactions between electrical, thermal, and mechanical forces that form, support, or break the flocs are very important for the understanding of the transport properties of nanofluids.

It must be noted that the Φ versus δ curve depicted in Fig. 4.4 is the result of several molecular, electrical, and hydrodynamic forces that depend not only on the properties of the particles but also on the characteristics of the base fluid. Significant modifications of these curves and abrupt changes in the relative magnitudes and positions of the energetic maximum and the two minima occur with simple modifications of the characteristics of the base fluid, such as changes of the solute molarity and pH. Such modifications have corresponding effects on the structure of nanofluids and their properties. Several of these modifications, the modeling of the attractive and repulsive forces during the flocculation process, and the stability of flocs are described in detail in specialized monographs on colloid substances, such as those by Russel et al. (1989), Probstein (1994), and Berg (2010).

4.4   Thermodynamic Properties

The equilibrium thermodynamic properties of a mixture are expressed in a simple way by the corresponding properties of the constituent materials using the theory of heterogeneous mixtures (Gibbs 1878). In this section we derive, as examples of the application of classical thermodynamic theory, three equilibrium thermodynamic properties: density, thermal expansion coefficient, and specific heat capacity. Based on these examples, one may derive all the other equilibrium thermodynamic properties for the heterogeneous mixture of the base fluid and the nanoparticles.

Let us consider a nanofluid composed of a number, n, of distinct particles enclosed by a volume V, as shown in Fig. 4.5. The fluid matrix occupies the volume V_f and the n nanoparticles occupy the volume V_s, which is equal to the sum of the volumes of the individual particles. The size of the volume, V, is large enough to satisfy the conditions for the fluid to be considered as a continuum according to the continuum theory described in Sect. 4.2.

$$V_s = \sum_{i=1}^{i=n} V_i. \tag{4.21}$$

Fig. 4.5 Nanofluid
composition

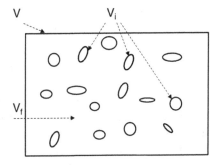

From the principle of volume conservation, we also have the equation $V = V_s +$
V_f. We define the volume fraction, ϕ, of the solids as the ratio V_s/V. It follows from
the volume conservation principle that the volume fraction of the fluid is $V_f/V = 1$
$- \phi$. The mass of the material enclosed in the volume V may be expressed in terms
of the bulk densities of the solid and fluid materials that comprise the nanofluid,
ρ_s and ρ_f, respectively. The total mass is:

$$m = [\rho_s\phi + \rho_f(1 - \phi)]V. \tag{4.22}$$

Hence, the average density of the mixture, which is simply defined as the ratio of
the mass of the mixture to its volume, is:

$$\rho_m = \frac{m}{V} = \rho_s\phi + \rho_f(1 - \phi). \tag{4.23}$$

Therefore, the density function is the volume-weighted average of the bulk
densities of the constituent substances of the nanofluid.

Another useful property of nanofluids is the thermal expansion coefficient, β,
which is defined as:

$$\beta = \frac{1}{v}\left(\frac{\partial v}{\partial T}\right)_P = \frac{1}{V}\left(\frac{\partial V}{\partial T}\right)_P = -\frac{1}{\rho_m}\left(\frac{\partial \rho_m}{\partial T}\right)_P. \tag{4.24}$$

The two constituent materials of the mixture, fluid and solid, have different
expansion coefficients, β_f and β_s, respectively. A change of the temperature of the
mixture, δT, at constant pressure will result in the following change of the volumes
of the two materials, which expand independently:

$$\delta V_s = \beta_s V_s\delta T = \beta_s V\phi\delta T \text{ and } \delta V_f = \beta_{fs}V_f\delta T = \beta_f V(1 - \phi)\delta T. \tag{4.25}$$

From the principle of volume conservation, the total volume change of the
mixture is equal to the sum of the volume changes of its constituents. For a mixture

with no mass exchange between the components, this yields the following for the
change of volume and an expression for the expansion coefficient, β:

$$\delta V = [\beta_s \phi + \beta_f (1 - \phi)] V \, \delta T \Rightarrow \beta = \frac{1}{V} \left(\frac{\partial V}{\partial T}\right)_P = \beta_s \phi + \beta_f (1 - \phi). \quad (4.26)$$

The last equation shows that the expansion coefficient of the mixture is given
simply as the volumetrically weighted average of the corresponding properties of
the constituent materials of the mixture. This rigorous derivation of the expansion
coefficient shows that certain "models," which have defined the property β as a
specific thermodynamic property in terms of the mass fraction of the heterogeneous
mixture, are incorrect.

An interesting result for the dependence of the volume fraction, ϕ, on the
temperature of the mixture may be obtained by deriving the expansion coefficient,
β, through the differentiation of the density function of the mixture, ρ_m, which is
given by Eq. (4.23).

$$\beta = -\frac{1}{\rho_m} \left(\frac{\partial \rho_m}{\partial T}\right)_P = \frac{1}{\rho_m} \left[\rho_s \phi \beta_s + \rho_f (1 - \phi)\beta_f - (\rho_s - \rho_f)\left(\frac{\partial \phi}{\partial T}\right)_P\right]. \quad (4.27)$$

A combination of the last two equations produces the following expression for
the variation of the volume fraction with the temperature:

$$\left(\frac{\partial \phi}{\partial T}\right)_P = \frac{1}{\rho_f - \rho_s} [\phi \beta_s (\rho_m - \rho_s) + (1 - \phi)\beta_f(\rho_m - \rho_f)]. \quad (4.28)$$

This yields the interesting result that the volumetric composition of the heterogeneous mixture changes with the temperature, under constant pressure.

A thermodynamic property such as the specific heat capacity, c, of the nanofluid
is based on the mass and not the volume of the two constituent phases and may not
be simply given as the volumetric-weighted average of the specific heat capacities
of the constituents. Referring to the constant mass heterogeneous thermodynamic
system of Fig. 4.5, the enthalpy of the mixture and the specific heat of the mixture
may be given as follows:

$$H = m_s h_s + m_f h_f \text{ and } c = \frac{1}{m} \left(\frac{\partial H}{\partial T}\right)_P = \frac{1}{m_s + m_f} (m_s c_s + m_f c_f). \quad (4.29)$$

It is implicitly assumed in the last expression that the addition of the
nanoparticles to the base fluid does not involve any phase change, chemical
reaction, energy release, or energy absorption. Expressing the mass of the fluid
and the solid phase in terms of their bulk densities and the corresponding volume
fractions yields the following expression for the specific heat capacity of the
heterogeneous mixture in terms of the volumetric fraction, ϕ:

$$c_m = \frac{1}{\rho_m} [\phi \rho_s c_s + (1 - \phi)\rho_f c_f], \quad (4.30)$$

which signifies that the specific heat capacity of the nanofluid is the mass-weighted average of the specific heat capacities of the constituent materials. Other thermodynamic functions of nanofluids, such as the entropy and all forms of energy, have a similar dependence on the corresponding properties of the constituent materials.

It must be noted that a few authors have defined a volume-averaged specific heat $[(1 - \phi)c_f + \phi c_s]$. This expression appears occasionally in the literature as another "model" for the specific heat capacity (e.g., *model I* in O'Hanley et al. 2012, and *model II* in Khanafer and Vafai 2011). Equation (4.30) is sometimes called an alternative model. However, it must be emphasized that the specific heat capacity is a well-defined thermodynamic property, which is defined by Eq. (4.29), and that there is no physical basis for any other "model" or definition of this thermodynamic property. Any other model or definition defines another function, e.g., a volumetric heat capacity, which is entirely different from the thermodynamic property that has been called *specific heat capacity* and has been used to design thermal systems for almost two centuries. This is corroborated by all the available experimental data, which show that Eq. (4.30) predicts very well the experimentally determined specific heat capacity of nanofluids while the other definitions or "models" do not (O'Hanley et al. 2012; Khanafer and Vafai 2011).

4.5 Transport Properties

Unlike the equilibrium thermodynamic properties, the transport properties of heterogeneous mixtures, such as the viscosity, the thermal conductivity, and the diffusivity, are defined by local gradients. There is not an established theory or scientific method to derive the transport properties of heterogeneous mixtures in terms of the corresponding equilibrium bulk properties of the constituent materials. Because of this, scientists have to rely on experimental measurements for the determination of the transport properties of nanofluids.

A great deal of research on the production, the characteristics, and properties of nanofluids stems from early experimental results, which indicated that the addition of a small fraction (less than 1%) of certain types of nanoparticles in a base fluid (typically water) increased significantly, sometimes doubled, the thermal conductivity of the mixture. It is well known that the current barrier to the further miniaturization of microelectronic devices and computer processors is the removal of the energy that is dissipated as heat. As the computer components become smaller, more heat is generated per unit area of the processors, which must be removed before the materials overheat and lose their semiconductor properties. Typical electronic components lose their semiconductor properties and malfunction above 120°C. Conventional cooling methods are not capable to remove effectively the dissipated heat from such miniature electronic devices without exceeding this temperature. The discovery that the nanofluids exhibit significantly higher thermal conductivity than the corresponding base fluids holds the promise that specifically produced nanofluid materials may be used as the cooling media of the next generation of electronic components. Several other industrial processes may also be optimized using nanofluid heat transfer.

The viscosity and thermal conductivity are two transport properties that determine the heat transfer and pumping characteristics of nanofluids. The functional form and variables that influence these two transport properties will be examined in detail in the next two sections.

4.5.1 Viscosity of Nanofluids

Viscosity is the resistance of fluids to motion. The dynamic viscosity of a Newtonian fluid is defined in terms of the local shear stress that is developed by the motion of the fluid and the local velocity gradient as follows:

$$\mu = \frac{\tau}{(\partial u / \partial y)}. \tag{4.31}$$

The kinematic viscosity of a fluid, v, is equal to the ratio μ/ρ_f. Equation (4.31) applies at all points of a fluid in motion. Solid materials do not flow at the application of shear stresses below their strength limit. For all practical purposes, solids exhibit infinite viscosity. The dynamic viscosity of a homogeneous, Newtonian fluid, μ, is a material property of the fluid. It depends only on the temperature and the pressure and is independent of the rate of shear $(\partial u / \partial y)$. Actually, the dynamic viscosity of a liquid is a strong function of the temperature and a weaker function of the pressure. All observations have shown that the viscosity of homogeneous fluids is constant within a fluid region where T and P are constant.

The dynamic viscosity of homogeneous fluids is typically measured by a rotating disk, a rotating cone, or a rotating sphere. Analytical solutions of the flow field and the stresses developed by these rotating objects have been developed in the past (Kestin et al. 1980). The viscosity of the fluid retards the motion of the rotating object in the fluid, and the measurement of the retardation yields a value for the viscosity. Very accurate measurements of the viscosity of homogeneous fluids, with accuracies of the order of 0.5%, have been accomplished using these rotating solid objects.

A simpler, albeit less accurate, method to measure the viscosity is to let the fluid flow in a laminar regime through a small, thin, or capillary tube under a known pressure, ΔP. The velocity profile in the capillary attains the typical quadratic profile of laminar flow. An integration of the Poiseuille velocity profile yields the viscosity of the fluid as a function of the volumetric flow rate, $\dot{V}$; the length of the tube, L; and the diameter, D, as follows:

$$\mu = \frac{\pi \Delta P D^4}{128 \dot{V} L}. \tag{4.32}$$

Fig. 4.6 Nanofluid with solids in two different configurations at the same volumetric fraction, ϕ. Case (b) exhibits higher viscosity than case (a)

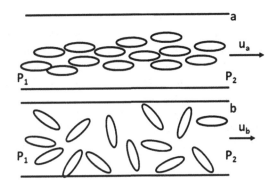

In the case of heterogeneous mixtures, such as nanofluids, the property of viscosity is not as easily defined, because the viscosity of a liquid–solid mixture cannot be defined as a point function in regions occupied by the solid phase. The solid particles move with uniform velocity and this does not allow Eq. (4.31) to be applied to all points of the flow domain. One may bypass this difficulty and define the viscosity function at the points occupied by the fluid alone, where Eq. (4.31) is meaningful. A moment's reflection though will prove that a viscosity defined in this way will be a strong function of the distribution of the solids inside the fluid because of the following:

1. Particles create a secondary velocity profile around them, which depends on the particle relative velocity and the distance from the surface of the particle.
2. Particle interactions, which involve both hydrodynamic and electrical interactions, create high fluid velocity gradients between them.
3. Particle collisions with the walls are inelastic and create additional stresses that contribute to the local fluid stress.

For the process of cooling of electronic components, engineers are interested in the volumetric flow rate of nanofluid that passes through a channel. In this case one may adopt an operational definition of the viscosity of the heterogeneous mixture according to Eq. (4.32). Thus, one may define the *bulk viscosity* of the heterogeneous mixture in terms of the volumetric flow rate of the mixture that flows through a capillary tube at a given pressure difference, ΔP. This definition would be sufficient for the needs associated with the design of cooling channels for microelectronics. However, it is apparent that such an operational definition would depend strongly on the distribution of the solids within the fluid matrix and the relative motion and migration of the particles. Figure 4.6 shows two particle configurations of the flow of a nanofluid in a narrow channel, both with the same ϕ. Direct numerical simulations as well as simple visual observations prove that under the same pressure drop, $\Delta P = P_1 - P_2$, the average mixture velocity in configuration (a) would be higher than that of configuration (b) and that $u_a > u_b$. The operational definition of viscosity of Eq. (4.32) would imply that $\mu_a < \mu_b$, despite the fact that the two heterogeneous mixtures have the same volume fraction and mass fraction of solids.

It must be noted that the other viscometers, which were designed and calibrated for homogeneous fluids (e.g., rotating disk, rotating cone, or rotating sphere) would also give different values for the viscosities when the structure of the solid particles is different. Actually, the rotating viscometers may give erroneous results for heterogeneous mixtures of solid particles if the particles or aggregates of particles are trapped in the narrow parts of the instruments. It is apparent from all measurements and experimental techniques that any operational definition of the viscosity of a nanofluid would depend greatly on the structure/configuration of the solid particles as well as on the properties of the fluid and solids.

4.5.1.1 Analytical Expressions and Correlations

The analytical investigations on the viscosity of fluids with solid spheres commenced with the study of Einstein (1906) who performed a first-order asymptotic analysis and determined that the viscosity of dilute ($\phi < 0.05$) heterogeneous mixtures with spherical particles is approximately equal to:

$$\mu = \mu_f(1 + 2.5\phi). \tag{4.33}$$

Brinkman (1952), Krieger (1959), Frenkel and Acrivos (1967), and Lundgren (1972) extended this form of analysis to slightly denser concentrations of spheres and derived similar, albeit more complex, expressions for the viscosity of a particulate mixture. The assumptions in all these investigations are that the solid particles are geometric spheres and electrically neutral and they do not aggregate. Brownian motion and its effects have also been neglected in these studies. Batchelor (1977) performed an analytical study that took into account the effects of the Brownian motion in a stochastic way, as well as the interactions of pairs of spherical particles, and derived an expression, which is applicable to very fine particulates at concentrations up to $\phi = 0.1$:

$$\mu = \mu_f(1 + 2.5\phi + 6.5\phi^2). \tag{4.34}$$

A glance at the last two equations proves that the accounting for pairwise interactions and the Brownian motion analytically makes a very small difference in the final result. For example, at the highest volumetric ratio Eq. (4.34) is normally valid ($\phi = 0.1$), the last term in the parenthesis accounts for less than 5% of the calculated viscosity value.

Early experimental results of the viscosity of particulates and especially the experimental data for nanofluids have shown that the viscosity values of liquids with nanoparticles are significantly higher than those predicted by the Einstein or the Batchelor equations. Other experimental studies agreed with the analytical results of the past. For example, Wang et al. (1999) suggested the following correlation for the viscosity of nanofluids:

$$\mu = \mu_f(1 + 7.3\phi + 123\phi^2), \tag{4.35}$$

while Chen et al. (2007) concluded from several sets of data that Batchelor's (1977) expression, Eq. (4.34), adequately correlates their measurements.

Other authors, including Mooney (1951), Tseng and Lin (2003), Tseng and Chen (2003), and Nguyen et al. (2007), determined that the viscosity of liquid suspensions is better correlated with exponential functions of the form

$$\mu = A\mu_f \exp(B\phi).$$ (4.36)

The values of the coefficients A and B vary significantly in these studies. Most of the investigations suggest that the coefficients depend on the type of solid particles used in the pertinent studies. Some investigators have also suggested that different correlations apply to the same materials but with different particles sizes.

Masoumi et al. (2009) performed an analytical study and incorporated the effects of the Brownian motion of particles in a simple model for the additional stress induced by the particles on the heterogeneous mixture. They derived the following expression for the effective viscosity of the mixture:

$$\mu = \mu_f \left(1 + \frac{\rho_s}{24C} \sqrt{\frac{k_B T}{\pi \rho_s \alpha}} \sqrt[3]{\frac{6\phi}{\pi}} \right),$$ (4.37)

where C is an empirical constant that was correlated by the available experimental data in terms of the volumetric ratio, ϕ; the diameter of the nanoparticles, 2α; and the temperature, T.

Most of the experiments with nanofluids have shown that the viscosity of the mixture is significantly higher than the viscosity of the base fluid and also higher than the predictions of all the analytical studies that treat the particles as inert solid spheres. Several measurements have demonstrated that the viscosity of the mixture doubles with the addition of a small fraction of nanoparticles ($\phi < 0.15$), while some studies showed a higher effect, where the viscosity of the mixture quadrupled or quintupled (Anoop et al. 2009; Chen et al. 2007). However, there has not been a general study or empirical correlation that applies to several types of base fluids and nanoparticles. Given the number of variables involved and the complexity of the subject, and the evidence that viscosity depends on the spatial distribution of the particles, an accurate and general correlation for the viscosity of these heterogeneous mixtures may be impossible to derive.

Because of the wide range of the available experimental data and the apparent significant differences of the experimentally observed viscosity of different nanofluids, several researchers suggested specific correlations that are applicable to one type of nanofluid only and in specific ranges of the parameters of interest. For example, Khanafer and Vafai (2011) collected several sets of experimental data for nanofluids composed of water and alumina particles (Al_2O_3) and performed a specialized correlation of the mixture viscosity in terms of the parameters ϕ, α, and T. Given the lack of reliable analytical model that would encompass several types of base fluids and nanoparticles, this may be a practice that will continue in the near future with several other nanofluids. The Khanafer and Vafai (2011) correlation is in the form of a power series and has the form

$$\mu = C_1 + \frac{C_2}{T} + C_3\phi + C_4\phi^2 + \frac{C_5\phi^2}{T^2} + C_6\phi^3 + \frac{C_7\phi}{T^3} + \frac{C_8\phi^2}{(2\alpha)^2} + \frac{C_9\phi^3}{(2\alpha)^2}. \quad (4.38)$$

The correlation coefficients C_1 through C_9 have been determined by a linear regression of the experimental data and their values are

$C_1 = -0.4491$	$C_2 = 28.837$	$C_3 = 0.574$	$C_4 = -0.1634$	$C_5 = 23.053$
$C_6 = 0.0132$	$C_7 = -2,354.735$	$C_8 = 23.498$	$C_9 = -3.0185$	

This correlation has a regression coefficient $R^2 = 0.99$ and is applicable in the ranges $0.01 < \phi < 0.09$, $20°C < T < 70°C$, and 13 nm $< 2\alpha < 130$ nm. It pertains to aquatic mixtures of Al_2O_3 only.

4.5.1.2 Rheological Behavior of Suspensions

One of the reasons for the significantly different viscosity values in heterogeneous liquid–solid mixtures is the effect of the electrostatic forces on the distribution of particles within the base fluid. Electrostatic forces determine the configuration of the solid particles in the nanofluid and, in several cases, may "trap" parts of the fluid within the particle aggregates, which are formed by the flocculation of the particles. When a significantly high part of the fluid is trapped in the solid structure, the entire mixture behaves as a weak solid, which is often called a *gel*. When the solids are in solution, the mixture is called a *sol*. Experimental observations have shown that several simple nanofluids exhibit the rheological behavior of non-Newtonian fluids and often form gel-like mixtures. One of the salient characteristics of non-Newtonian fluids is that their apparent viscosity is not only a function of T and P but also a function of the local value of the shear, γ. This implies that the viscosity of non-Newtonian fluids is not a property of the material alone. For this reason, it is called effective viscosity and is denoted by the symbol η:

$$\eta = \frac{\tau}{\gamma}. \quad (4.39)$$

It is apparent that $\eta = \eta(T, P, \gamma)$. Figure 4.7 shows a typical dependence of the effective viscosity of a nanofluid on the rate of shear, γ, for constant values of T and P. The data in the figure are from Krieger (1972) and pertain to polystyrene nanoparticles with sizes in the range 108–180 nm in deionized water. The effective viscosity is made dimensionless by the dynamic viscosity of the base fluid, η/μ, and the shear is expressed in terms of the dimensionless shear, or shear Peclet number, $Pe = \mu\alpha^3\gamma/k_BT$. It is observed that the effective viscosity exhibits two asymptotic limits at low and high values of the shear, η_0, and η_∞. Both these asymptotic limits

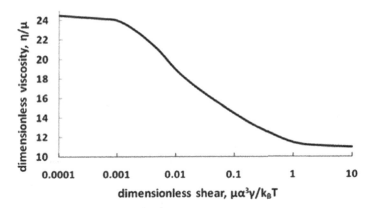

Fig. 4.7 Shear dependence of effective viscosity of a nanofluid. Data from Krieger (1972)

are significantly higher than the viscosity of the base fluid. The value of the effective viscosity at high shear is much lower than the effective viscosity at the lower shear rate. This behavior of the particulate mixtures to exhibit significantly lower effective viscosities at high shear rates is called *shear thinning*. The configuration of the particles within the base fluid matrix and the particle behavior and interactions during the process of shear are the main parameters that contribute to shear thinning (Russel et al. 1989; Berg 2010).

The formation of a weak gel structure and shear thinning behavior have been observed in several nanofluids. Kim et al. (2011) observed that nanofluids composed of deionized water and alumina nanoparticles exhibited the rheological characteristics of Newtonian fluids at low volume fractions ($\phi \leq 2\%$) but showed a strong non-Newtonian behavior, including shear thinning at higher volume fractions ($3\% \leq \phi \leq 5\%$). Two representative curves from the data of Kim et al. (2011) are shown in Fig. 4.8, where it may be seen that the dilute nanofluid with $\phi = 1\%$ exhibits the typical behavior of a Newtonian fluid and the ratio η/μ remains constant and approximately equal to 1 (within the experimental uncertainty of the data) at shear rates that span three decades. On the contrary, the nanofluid composed of the same materials with $\phi = 4\%$ exhibits strong non-Newtonian behavior with its effective viscosity η reaching values close to 100 times higher than the base fluid viscosity, μ, at the low shear rates. This nanofluid exhibits strong shear thinning behavior and its effective viscosity is reduced to 4μ at the higher rates of shear. Kim et al. (2011) observed that the particles in the nanofluids within the volumetric range $3\% \leq \phi \leq 5\%$ flocculated, and the entire solution formed a gel-like structure, which explained the very high values of η at the low shear rates.

Kim et al. (2011) also observed that the preparation process and time to settle of the nanofluid played a role in the value of its effective viscosity: the milling process of the nanoparticles and the amount of time the solution was left undisturbed had a significant effect on the characteristics and rheological behavior of the nanofluids. For example, a nanofluid with $\phi = 4\%$ was milled for 5 h and when tested immediately after, it showed Newtonian behavior. When the same nanofluid was

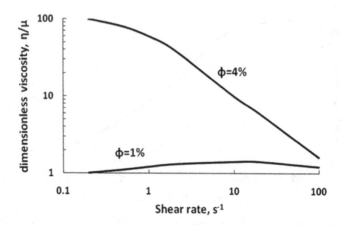

Fig. 4.8 Rheological behavior of a water–alumina nanofluid at two different solids concentrations. Data from Kim et al. (2011)

left for 6 days, it exhibited a strong non-Newtonian behavior with $\eta = 80 \ \mu$ at low shear rates. These experiments suggest that the particles in the water–alumina nanofluids flocculated and formed weak gels. The structure of the gels was broken by the external forces of the milling process but was restored when no forces were imposed for a prolonged time.

The electrostatic forces in the fluid–solids mixture are the main contributors to the spatial configuration of the particles, the flocculation process, and the formation of gels. Because of this, the pH of the mixture is a variable that affects significantly the values of η_0, and η_∞, the onset of shear thinning, and by extent the effective viscosity of the particulate mixture. This behavior has been shown in the early experiments by Krieger and Eguiluz (1976) and has been documented in several monographs (Russel et al. 1989; Berg 2010). The experiments by Kim et al. (2011) also corroborated this dependence: When the pH was increased from 4 to 7 in a nanofluid with $\phi = 5\%$, the effective viscosity increased by three to four orders of magnitude for the entire range of shear rates. In similar experiments, a dilute nanofluid with $\phi = 1\%$ exhibited gel-like structure and non-Newtonian behavior when the pH increased from 4 to 7.

4.5.2   Thermal Conductivity

The early experimental work of Choi et al. (2001) reported that the addition of a small fraction of carbon nanotubes in engine oil increases the thermal conductivity of the base fluid by a factor of 2–2.6. This implies 100–160% heat transfer enhancement. This enhancement was initially characterized as "anomalous" and sparked immense scientific interest on the thermal characteristics of all nanofluids during the first decade of the twenty-first century. Among the base fluids that have

Table 4.1 Thermal conductivities of several materials common in nanofluids

Solids	k_s (W/mK)	Liquids	k_f (W/mK)
Silver (Ag)	427	Water	0.613
Copper (Cu)	395	Ethylene glycol	0.253
Aluminum (Al)	237	Engine oil	0.145
Carbon nanotubes	3,200–3,500	Alcohol	0.115
Brass	120	Glycerol	0.285
Nickel	91		
Quartz (single crystal)	7–12		
Alumina (Al_2O_3)	39		

been investigated for the composition of nanofluids are water, engine oils, and ethyl glycol. Aluminum oxide (Al_2O_3), copper oxides (both CuO and Cu_2O), single-walled and multiwalled carbon nanotubes (CNT), copper (Cu), and gold (Au) nanoparticles are some of the solid particles that have been used in the experimental studies on nanofluids. Table 4.1 lists the thermal conductivities of several materials that are commonly used as base fluids and nanoparticles. It is observed that the thermal conductivities of the solid particles are several orders of magnitude higher than the conductivities of the base fluids.

The results for the heat transfer enhancement in the several experimental studies show large variations: While most of the investigators observed an enhancement of the thermal conductivity of the suspensions, this enhancement was typically in the range 5–50% and did not approach the high levels of the study by Choi et al. (2001). Only one other study by Chopkar et al. (2008) reported enhancements close to 100% in a suspension of Ag_2Al particles in water and ethyl glycol base fluids. A review by Khanafer and Vafai (2011) pertaining to Al_2O_3 particles in water and ethyl glycol showed that typical conductivity enhancements are in the range 4–30%, even when the volumetric fraction of solids is relatively high, close to 6%. From the several experimental studies, it appears that the most important parameters of conductivity enhancement in nanofluids are the type and properties of nanoparticles, the type of the base fluid, and the volumetric fraction of the solids in the nanofluid (Khanafer and Vafai 2011; Kakaç and Pramuanjaroenkij 2009).

4.5.2.1 Analytical Expressions and Correlations

Regarding the analytical work on this subject, an expression for the effective thermal conductivity, k_e, of a homogeneous mixture of spheres in a fluid may be obtained following the analogous derivation for the electrical conductivity by Maxwell (1881). For a fluid that has lower thermal conductivity than the solid particles ($k_f < k_s$), Maxwell's expression becomes

$$k_e = k_f \left[1 + \frac{3(k_s - k_f)\phi}{(k_s + 2k_f) - (k_s - k_f)\phi} \right]. \tag{4.40}$$

Similarly, one may use the analytical work by Bruggeman (1935) on the electrical conductivity of particles to derive the following expression for the thermal conductivity of a suspension of cylindrical or spherical particles:

$$k_e = k_f \left[\frac{k_s + k_f(n-1) + (n-1)(k_s - k_f)\phi}{k_s + k_f(n-1) - (k_s - k_f)\phi} \right], \tag{4.41}$$

where the shape factor $n = 3$ for spheres and $n = 6$ for cylinders.

Hamilton and Crosser (1962) derived a model for the conductivity of liquid–solid suspensions of irregular particles and introduced a shape factor to account for the influence of the shape of particles. Bonnecaze and Brady (1990, 1991) developed a theoretical framework for the computation of the thermal conductivity of suspensions. A subsequent study by Nan et al. (1997) introduced a methodology for the calculation of the effective thermal conductivity of composite materials taking into account the shape, symmetry, and orientation of the particles. Of particular interest is that the theoretical expression of Nan et al. (1997) applies to fibers of very long aspect ratio, such as carbon nanotubes. The theory predicts that when long fibers of solid materials with high conductivity are used with liquid coolants, the effective conductivity of the heterogeneous mixture increases significantly. Nan et al. (2003) developed a simplified model for nanotubes of high aspect ratio and with conductivities much larger than that of the base fluid ($k_f \ll k_s$). The approximate expression obtained by Nan et al. (2003) may be written as follows:

$$k_e = \frac{3k_f + \phi k_s}{3 - 2\phi} \approx k_f + \frac{\phi k_s}{3}. \tag{4.42}$$

According to this simplified theory, the addition of very small amounts of single-walled or multiwalled carbon nanotubes—materials that have thermal conductivities of the order of 1,000 W/mK—may triple or quadruple the thermal conductivity of the mixture even at volumetric concentrations close to 1%.

Among the expressions for the enhanced conductivity of nanofluids, a recent study by Khanafer and Vafai (2011) used several sets of data to derive a correlation for the effective conductivity of suspensions of Al_2O_3 and CuO nanoparticles in water at ambient temperatures:

$$\frac{k_e}{k_f} = 1 + 1.0112\phi + 2.4375\phi \frac{47}{d_s} - 0.0248\phi \frac{k_s}{k_w}, \tag{4.43}$$

where d_s is the particle size in nm and k_w is the thermal conductivity of water, $k_w = 0.613$ W/mK. Because this is strictly a correlation of experimental data, it should not be used outside the range of its applicability. One of the shortcomings of this correlation—and for which there is no physical explanation—is why the conductivity of the solids affects adversely the conductivity of the suspension (the pertinent coefficient of the correlation is -0.0248).

The very high values of thermal conductivity observed in the early experiments have led several authors to characterize the enhancement of the thermal conductivity of nanofluids as "anomalous." These enhancements were of the order of 150% and were observed primarily with carbon nanotubes that have very high aspect ratios (100–1,000) and very high conductivities, of the order of 1,000 times higher than the conductivity of water. Such highly elongated particles are covered by expressions similar to Eq. (4.42), which also predicts very high enhancements with highly conducting nanoparticles. A benchmark study by Buongiorno et al. (2009), which was conducted with the participation of 34 laboratories worldwide, used the applicable analytical expressions to compare several sets of experimental data and concluded that the observed heat transfer enhancement with nanofluids is not "anomalous" and that the experimental data are well explained by one of the pertinent equations (4.40), (4.41), or (4.42).

4.5.2.2 Mechanisms for the Increased Conductivity

The following mechanisms have been suggested for the enhanced conductivity of nanofluids:

1. The higher heat conductivity of the particles
2. The formation of aggregates that form or enhance highly conductive paths in the nanofluid
3. Significant changes of the thermodynamic properties of the fluid at the solid–fluid interface and the formation of a "solid layer"
4. The electric charge on the surface of the particles
5. The Brownian motion of the particles, which also includes the fluid that follows the motion of the particles
6. Transient local heat transfer effects

The higher conductivity of the dispersed solids increases the heat conductivity of the suspension because the solids offer pathways to conduct heat more readily. This is a fact that has been observed in most of the experimental and analytical studies, regardless of the shape of the particles. In the case of elongated solids (nanotubes, fibers, and cylinders), the conductivity enhancement is higher because the pathways are longer and capable to conduct heat further (Choi et al. 2001; Nan et al. 2003).

As with the viscosity of nanofluids, the spatial distribution of the particles and the formation of aggregates would also influence the thermal conductivity of the suspension. Particles or aggregates of particles may form structures and layers that would significantly enhance the conductivity of the suspension, even at the same volumetric concentration. This is depicted schematically in Fig. 4.9. From visual observations, it may be deduced from the figure that the effective thermal conductivity of arrangement (b) would be significantly higher than that of arrangement (a) because the ellipsoidal particles in arrangement (b) have aggregated in chains to form several highly conducting paths. Several experimental and numerical studies on particle aggregation and the formation of highly conducting paths support

Fig. 4.9 The particulate
structures and formation of
aggregates and particle chains
have a significant effect on
the thermal conductivity. The
effective conductivity of
arrangement (b) is higher than
that of (a)

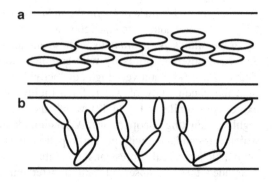

this mechanism (Prasher et al. 2006b; Timofeeva et al. 2007; Philip et al. 2008). In particular, the study by Timofeeva et al. (2007) who conducted simultaneously conductivity and particle size measurements concludes that the measured conductivity is significantly affected by the size of the aggregates in the nanofluid. Given that the aggregation, flocculation, and separation processes are dynamic processes that continuously change the particulate structures in a nanofluid, this would also imply that the thermal conductivity varies accordingly. These studies also point to the fact that accurate measurements of the thermal conductivity of the nanofluid must be accompanied by the knowledge of the particle distribution or at least of the sizes of the particle aggregates.

The mechanism of the formation of a solid layer (liquid layering) at the interface was proposed by Choi et al. (2001) and Keblinski et al. (2002). Xue (2003) developed a model, based on this layer and Maxwell's theory of electrical conductivity, which explained the observed increase of k_e. Soon thereafter, Xue et al. (2004) extended this model using molecular dynamics and concluded that "... the experimentally observed large enhancement of thermal conductivity in suspensions of solid nanosized particles (nanofluids) cannot be explained by altered thermal transport properties of the layered liquid." Among the experimental studies of this mechanism was the one by Shin and Banerjee (2010, 2011) who attributed the increased specific heat capacity and part of the conductivity enhancement of nanofluids on the formation of a solid layer around the particles. However, if such a layer were formed, its width would be of the order of a few molecular dimensions, that is, of the order of 1 nm. The volume and mass of the fluid that is immobilized or "solidified" on the surface of the nanoparticles is very small in comparison to the volume of the particle. Nie et al. (2008) showed analytically that the immobilized fluid volume is simply too small to account solely for the observed change in the conductivity of nanofluids. It is highly likely that the observed "solid layer" is actually the electrostatic double layer, which was described in Sect. 4.3.4 and is composed of the base fluid in the liquid state.

As with the viscosity of nanofluids, the electric charge on the surface of the particles and the zeta potential affect the structure of the solids in the nanofluid and by extent the thermal conductivity. Lee et al. (2006) investigated experimentally the influence of the surface charge of nanoparticles on the thermal conductivity of the suspension and concluded that the surface charge has significant effects on the effective conductivity. They concluded that departures from electrical neutrality for the nanoparticles cause higher stability of the nanoparticle clusters and higher stability of the suspension fluid overall. This in turn enhances the thermal conductivity of the suspension. The zeta potential effect may partly explain the disparities between some sets of experimental data, where researchers used surfactants that alter the surface charge of the nanoparticles and, hence, the structure of the suspension. Jung and Yoo (2009) came to a similar conclusion after studying the effect of the electric double layer on nanoparticles. Related to these studies on the electric double layer are the ones by Wamkam et al. (2011) who examined the effect of the liquid pH of the base fluid and found significant changes in the conductivity of the suspension and by Fan and Wang (2011) who made a structure—property correlation for solid particles and clusters in liquids.

Since Brownian motion is the primary mechanism for the mechanical agitation in nanofluids—a mechanism that contributes to higher heat transfer—several studies have been conducted to quantify the effect of this motion on the thermal conductivity of nanofluids. Among the earlier analytical studies, Koo and Kleinstreuter (2005a, b) examined the effects of Brownian motion, the thermophoretic,[2] and osmophoretic motions on the effective thermal conductivities of nanofluids. They concluded that the role of the Brownian motion is the main mechanism for the observed enhanced conductivity and that it is by far more important than the thermophoretic and osmophoretic motions. They also confirmed a result, which is well known for larger particles: Particulate interactions have a negligible effect on the heat transfer coefficient of nanofluids when the concentration is very low. Koo and Kleinstreuter (2004) also proposed a model for the enhancement of the effective conductivity of nanofluids, based on the Brownian motion. They demonstrated the monotonic dependence of thermal conductivity with temperature with the help of this model. Kumar et al. (2004) also developed an analytical model, based on the Brownian motion that explained the dependence of the effective conductivity on the temperature.

An analytical study by Prasher et al. (2005) also concluded that the Brownian movement of particles is the main reason for the observed conductivity enhancement in nanofluids. It must be noted, however, that the analytical results of this study were confirmed with experiments on alumina nanofluids only, which show the least amount of conductivity enhancement. Prasher et al. (2006a) showed that the micro-convection caused by the Brownian movement of nanoparticles is a possible mechanism for the enhancement of the effective thermal conductivity of nanofluids.

Among the more recent studies, Yang (2008) used a kinetic theory approach and investigated analytically the effect of the Brownian motion on the thermal conductivity of nanoparticles. He concluded analytically that the Brownian motion plays

---

[2] It must be recalled from Sect. 4.3.3 that thermophoresis is the manifestation of Brownian motion in a temperature gradient. The two are not independent mechanisms to be treated independently.

an important role in enhancing the conductivity of the suspension and supported these results with comparisons with experimental data for the Al_2O_3-in-water nanofluids. Independently, Shukla and Dhir (2008) developed a simple model for predicting the thermal conductivity of nanofluids based on the Brownian motion. They examined the instabilities during the natural convection of nanofluids and concluded that, as a result of the Brownian motion and thermophoresis, the critical Rayleigh number, Ra_{crit}, for the inception of natural convection is significantly lower than that of the base fluid. This study also shows that the reduction of Ra_{crit} is one to two orders of magnitude with the addition of nanoparticles.

Not all the studies on the effect of Brownian motion in the effective conductivity of nanofluids agree. For example, the study by Evans et al. (2006) in stagnant fluids suggests that the contribution of Brownian motion to the thermal conductivity of a nanofluid is smaller and may not be solely responsible for any extraordinary thermal transport. Nie et al. (2008) concluded analytically that even though the Brownian motion of the nanoparticles enhances the heat transfer characteristics of nanofluids, this motion alone may not cause the very high conductivity enhancement that was observed in several studies. However, one of the shortcomings of this study is that it did not take into account the motion of the fluid that accompanies the particle motion. Similarly, the experimental study by Gao et al. (2009) which used a solid matrix of fat to "freeze" the Brownian movement of particles concluded that this motion is not the main cause for the increased conductivity of nanofluids. Also, Eapen et al. (2010) used an analytical argument to show that the "micro-convection" due to the Brownian motion would not affect significantly the thermal conductivity of a nanofluid. These conclusions imply that other, perhaps unknown, mechanisms are at play in nanofluids that need to be investigated.

It is apparent that the several analytical studies on the effect of Brownian motion on the thermal conductivity of nanofluids have resulted in an open disagreement and that the experimental data on the subject are not convincing. This significant disagreement may only be solved with definitive experiments and acceptable analyses, which will take into account not only the motion of the solid particles but also the induced movement of the fluid and all the associated effects in the bulk fluid. Accurate numerical simulations, perhaps stochastic simulations that take into account the randomness of the Brownian motion, would be helpful to clarify this subject.

Related to the Brownian motion are the effects of the transient motion and heat transfer from particles. Three such effects are dominant:

(a) The virtual or added mass of the base fluid that accompanies the motion of the particles
(b) The mass of the fluid that rushes to replace the volume of the moving particles
(c) The contribution of the history term on the heat transfer from the particles

The virtual mass of the base fluid that follows the transient motion of a spherical particle is equal to the fluid mass that occupies 1/2 the volume of the sphere and appears as the first term on the r.h.s. of Eq. (1.45). A similar amount of fluid follows particles of other shapes. As the particles move almost randomly, this mass of the fluid is transported to regions of different temperatures and contributes to the overall energy exchange. Similarly, when the particle and its surrounding fluid move, fluid from a different part of the suspension rushes to fill the void. The history term of the transient energy equation is the last term of Eq. (1.45). This term represents the effect of the changing temperature gradients in the fluid caused by the changes in the temperature of the particles (Michaelides and Feng 1994). As the Brownian motion carries the particles, the local fluid temperature gradients change and this process induces higher heat transfer from the particles to the fluid. The three transient effects result in the higher "agitation" of the base fluid and contribute to the heat transfer enhancement.

The Brownian motion is due to the molecular motion but manifests itself through the motion of the particles that have inertia; its effects would be evident at timescales that span the range from the particle characteristic timescales, τ_M and τ_{th}, to the much shorter molecular timescales. Hence, the Brownian movement of particles and the associated transient particle and fluid movement would have effects (a) on the thermal conductivity of the fluid, which is associated with the molecular momentum and energy transfer at the molecular timescales, and (b) on the convective heat transfer coefficient, which is associated with the advective motion and heat transfer and which occurs at the two particle timescales, τ_M and τ_{th}. The latter effect is examined in more detail in the next section.

4.5.3 Heat Transfer Coefficients

4.5.3.1 Convective Heat Transfer Coefficient

While the thermal conductivity is a good indicator of enhanced heat transfer in thermal systems, the convective heat transfer coefficient, h_c, is the parameter that best characterizes the performance of a fluid as a heat transfer medium. The few experimental studies that measured directly the convective heat transfer coefficient of nanofluids typically found that h_c was significantly higher than that of the base fluid under the same flow conditions. This may be partly explained because the effective conductivity of the nanofluids is higher than that of the base fluid. However, there is convincing experimental evidence that the enhancement of the convective heat transfer coefficient is actually higher than the enhancement of thermal conductivity. For example, the experimental study by Wen and Ding (2004), which pertained to dilute alumina suspensions in water, observed heat transfer coefficient enhancements between 35% and 45% at several values of Reynolds numbers with $\phi = 1.6\%$, when the measured conductivities of the nanofluid did not exceed 10%. At lower volumetric fractions, the heat transfer coefficient enhancement was lower, but consistently higher than the increase of the conductivity of the nanofluid.

Since advection effects are associated with the Reynolds number, rather surprisingly, the results by Wen and Ding (2004) do not show a clear enhancement of the

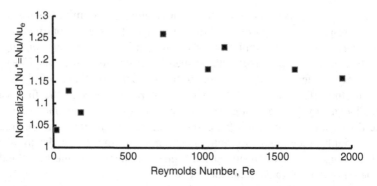

Fig. 4.10 Nusselt numbers, corrected for the enhanced conductivity of the nanofluid, versus Re. The data in the range $21 < Re < 210$ are from Lai et al. (2009) and the data with $Re > 700$ are from Wen and Ding (2004)

heat transfer coefficient with increasing Reynolds numbers. A subsequent study by Ding et al. (2006) with carbon nanotubes shows also a flat rate of enhancement in the range $800 < Re < 1,100$. The latter study only shows a consistent improvement of the Nusselt number with the Reynolds number only between $Re = 1,100$ and $Re = 1,200$, where Nu almost abruptly doubles when Re increases from 1,100 to 1,200. This suggests that other effects, such as entrance effects, an early flow transition to turbulence, or fluid agitation caused by particle inertia, may have played a role in the observed convective heat transfer enhancement. On the contrary, the results by Heris et al. (2006) show a more consistent trend of increasing Nu with increasing Re in the entire range of the data. Figure 4.10 depicts representative data of Nu versus Re taken from Lai et al. (2009) and Wen and Ding (2004), both for Al_2O_3 nanoparticles in water. The normalized Nusselt numbers, Nu*, are defined as the ratio of the actual Nu to the Nusselt number determined using the effective conductivity of the nanofluid and account for the increased conductivity of the nanofluid ($Nu_c = h_c L_{char}/k_e$). It is observed in this figure that, even though the trend Nu *versus* Re is not clear, all the data show consistently that the increase of the Nusselt numbers is significantly more than the effect of the increased conductivity of the nanofluid. This indicates that advective effects within the nanofluid, which include the motion of nanoparticles, play a significant role in the laminar convective heat transfer process.

Regarding nanoparticle types, it appears that the addition of metal oxide nanoparticles will result in heat transfer enhancements up to 50%. The highest enhancements of convective heat transfer coefficients, up to 150%, were observed in suspensions of carbon nanotubes in water (Ding et al. 2006), and this is consistent with the significantly higher effective conductivity enhancement of the nanofluid.

Only a limited number of experimental data are available on the turbulent flow regime for heat transfer coefficients with nanofluids. Most of the experiments lead to the same conclusions as the experiments with laminar flow: At all volume fractions, the increased heat transfer coefficients of nanofluid suspensions are higher than what the enhanced conductivity models and measurements would predict. For example, Torii and Yang (2009) measured heat transfer coefficient,

h_c, values 8–20% higher than what would have been calculated from the enhanced conductivity values ($h_c = k_e Nu/L_{char}$), in the range $3,500 < Re < 6,000$. The experimental results by Xuan and Li (2003) indicate very similar enhancement of h_c. On the other hand, a few of the experimental data by Pak and Cho (1998) show a decrease of the convective heat transfer coefficient. This may be due to an erroneous measurement of the viscosity that propagated to the calculation of Re. Another study by Williams et al. (2008) did not find significant deviations between their experimental data and the results from the Dittus–Boelter equation for turbulent flows, once the increased conductivity and viscosity were taken into account. In general, a review of the reported results on h_c suggests that more systematic research and more accurate experiments, based on sound measurement principles, are needed for the definitive characterization of the convective coefficients of nanofluids.

A possible explanation for the modification of the laminar convective coefficients is the influence of Brownian motion. While the Brownian movement of particles has its origins with the molecular collisions and its timescale is of the order of the molecular timescales, it is manifested through the particles that have inertia and respond at a much higher timescales, τ_M and τ_{th}. At these higher timescales, the movement of the particles agitates locally the fluid and this will show as increased h_c, not as increased k_e, which is usually measured by an asymptotic method at the inception of the heat transfer process and before any advection takes place. Parts of the transient effects that have been mentioned at the end of the previous section (Sect. 4.5.2.2) would also manifest themselves at the timescales that affect the convective heat transfer coefficient too and would contribute to the increase of h_c. Preliminary studies (Esparza 2012) and current numerical work by the author and his students at TCU indicate that the Brownian motion of the particles and the associated transient effects may fully account for the observed increase of the laminar convective heat transfer coefficient, h_c, of nanofluids at low solids volume fractions.

In the case of turbulent flow, the turbulent fluctuations, the turbulent motion of particles, and the resulting turbulent "agitation" are significantly stronger than the Brownian movement and the related fluid "agitation." For this reason, a significantly increased convective heat transfer coefficient, h_c, would not be supported by the "particle agitation" mechanism of the Brownian motion. It appears that this mechanism for enhance heat transfer coefficients is supported by all the studies on laminar flow with nanofluids and the experimental study by Williams et al. (2008) but not by the other studies on turbulent flow with nanofluids. Clearly, more research is needed on this mechanism to clarify the effects of the Brownian movement of particles and the associated transient effects on the heat transfer.

4.5.3.2 Macroscopic Results with Gas–Solid Suspensions

The enhanced heat transfer characteristics of fluid–solids suspensions is not a new phenomenon or research subject that started with nanofluids. High convective heat transfer coefficients of gas–solid suspensions have been observed during the 1960s, when gas–solid mixtures were considered as nuclear coolants. It has been well established by these early experiments that the presence of small and fine particles (50 μm–5 mm) significantly enhances the rate of heat transfer in laminar and turbulent duct flows of any cross section. Experimental studies of gas–solid flows (Schluderberg et al. 1961; Wachtell et al. 1961; Farbar and Depew 1963; Pfeffer et al. 1966) have shown that the addition of solid particles in gases enhances the convective heat transfer coefficient of the suspensions. This enhancement of the heat transfer coefficient is described by correlations of the experimental data, such as the one by Pfeffer et al. (1966):

$$\frac{h_c}{h_{c0}} = 1 + 4Re^{-0.32}m^* \frac{c_s}{c_f}, \tag{4.44}$$

where h_{c0} is the convective heat transfer coefficient of the gas, in the absence of particles; Re is the Reynolds number of the gas; and m^* is the loading, which is defined as the ratio of the solids-to-gas mass flow rates through the pipe. In nanofluids and flows with fine particles, $m^* \approx \phi \rho_s / \rho_f$.

The phenomenological model of particulate heat transfer by Michaelides (1986) explained the mechanism of the enhanced heat transfer in the gas–solid flows and showed that, in general, the presence of particles of any size in fluids enhances the energy transfer via the following mechanisms:

1. The particles in general have higher conductivity and heat capacity.
2. The particle motion agitates the fluid and causes velocity fluctuations. This was more recently called "pseudo-turbulence" (Lance and Bataille 1991; Lohse et al. 2004).
3. Local, microscopic energy exchange between particles and fluid.
4. In dense flows, local temperature gradients become sharper because of particle–particle and particle–wall interactions.

It is rather unfortunate that the research on the heat transfer with nanofluids, which occurred during the first decade of the twenty-first century, ignored the results of the significant research on gas–solids suspensions because most of the above mechanisms apply to nanofluids also. A preliminary study by Granger et al. (2012) suggests that the observed convective heat transfer enhancement with nanofluids is rather well correlated by the empirical expressions, which were developed in the 1960s for the gas–solids suspensions.

4.5.3.3 Pool Boiling and Critical Neat Flux

From the limited number of experimental studies on pool boiling, it appears that the presence of nanoparticles in liquids does not always increase the heat transfer coefficient. Results on heated tubes by Das et al. (2003) with Al_2O_3 particles in water showed a consistent trend of lower pool boiling coefficients, which was attributed to particle deposition on the heated tube surface. Bang and Chang (2005) did a similar study with Al_2O_3 nanoparticles in water and with a heated flat plate. Their results show a marked decrease of the rate of the pool boiling heat transfer coefficient with the addition of particles. The rate of heat transfer deteriorates at higher particle volume concentrations. On the contrary, Wen and Ding (2005) reported heat transfer enhancement in their Al_2O_3–water nanofluid, when the particle concentrations were less than 0.32%. The experimental study by Krishna et al. (2011) used a Cu–water nanofluid with a flat copper surface. Their results indicate that at low heat fluxes, there is a deterioration of the boiling heat transfer coefficients at the very low particle concentration of 0.01%. The pool boiling heat transfer coefficient increased with the concentration and, actually, became higher than that of water at $\phi > 0.1\%$. The situation was entirely different at the high heat fluxes, where an increase in ϕ caused the decrease of the pool boiling coefficient.

The heat transfer enhancement at low heat fluxes was attributed to the formation of a thin sorption layer of nanoparticles on the heater surface. The layer may trap some of the nucleation sites, but also it helps increase the micro-layer evaporation, which is due to the enhancement of the thermal conductivity of the fluid. The available experiments show that, in general, the addition of oxide particles decreases the pool boiling rate of heat transfer, while the addition of metals enhances the heat transfer. It also appears that nanoparticles enhance the heat transfer rate of pool boiling at lower heat fluxes and decrease it at higher fluxes. The following mechanisms may account for the observed trends on pool boiling heat transfer in nanofluids:

(a) Nanoparticles not only provide nucleation sites at low concentrations but also cover a fraction of the nucleation sites at higher concentrations.
(b) Nanoparticles deposit on the heating surface, thus altering the heat transfer characteristics of the surface.
(c) Nanoparticles alter the surface tension of the base fluid.
(d) Nanoparticles enhance the micro-layer evaporation.

The critical heat flux (CHF) occurs when the vapor produced at the heating surface forms a vapor layer that blankets this surface. The lower thermal conductivity of the vapor increases significantly the local heat resistance of the fluid. In order to maintain constant heat flux to the bulk of the fluid, the temperature of the heating surface rises. This rise of temperature can be significant and may have detrimental effects on the materials of the surface and the mechanical integrity of the heating system. For this reason, the operating conditions of thermal systems are far away from the CHF conditions. Increased CHF of cooling fluids is very welcome in thermal systems design, because it provides a higher safety margin.

All the currently available experimental data indicate that the addition of nanoparticles increases the CHF of base fluids, often by a factor of 2–3. You et al. (2003) observed with a horizontal heater a 200% increase of the CHF in water with the addition of trace amounts of Al_2O_3 nanoparticles $10^{-6} < \phi < 10^{-5}$. They also observed that the vapor bubbles departing from the surface were significantly larger in the nanofluid experiments than in the pure water experiments. This implies that the addition of nanoparticles affected the surface tension of the water. On the other hand, Bang and Chang (2005) observed only a 50% increase of

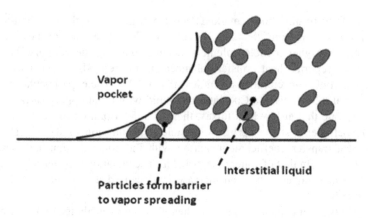

Vapor
pocket

Interstitial liquid

Particles form barrier
to vapor spreading

Fig. 4.11 Nanoparticles prevent the spreading of vapor patches on heated surfaces

the CHF, also with a horizontal heater, but with significantly higher volumetric fractions of nanoparticles $5 \times 10^{-3} < \phi < 5 \times 10^{-2}$.

CHF occurs because of the expansion and spreading of dry patches on the heating surface, which creates thin vapor layers. Since the particles do not evaporate with the fluid, they remain in the liquid layer. This creates higher particle concentrations adjacent to the dry patches and has two effects, both of which are adverse to the spreading of the dry vapor pockets:

(a) Restricts the sideways expansion of the vapor patch (Wen 2008)
(b) Maintains through surface tension interstitial liquid layers between individual particles and between the particles and the heated surface.

Figure 4.11 shows the effect of the addition of particles on the stabilization and spreading prevention of a vapor pocket under boiling conditions. It follows that irregularly shaped particles that may be "trapped" on the surface or aggregate to form an impediment to the lateral spreading of the vapor have a better effect on the CHF enhancement.

In general the addition of nanoparticles has an unresolved effect on the boiling heat transfer coefficient. However, it always increases the heat transfer flux as it appears in Huang et al. (2011). Given that the boiling heat transfer coefficients are very high anyway, their lower values will not have a significant effect on the design of most thermal systems. The increased CHF is significant though because it provides a higher margin of safety for boiling systems, which may be designed to operate at higher heat fluxes.

4.6 Concluding Remarks

The subject of heat transfer in nanofluids is relatively recent and too young for many definitive conclusions to be drawn. Even though the pertinent physical mechanisms have not been fully investigated, it appears that the addition of small amounts of particles enhances the conductivity of the base fluid and that there is an additional enhancement of the convective heat transfer coefficient. Also, it appears that there is an enhancement of the CHF. Clearly more investigations are needed to determine precisely the type and composition of heterogeneous mixtures that result in better heat transfer media. However, it is also clear that there needs to be a systematic investigation of the subject with well-defined and accepted measurement protocols and measurement methods. Experimental studies that simply report partial results without following established protocols, without significant supporting evidence on the experimental conditions, and without examination of mechanisms are not helpful.

One of the impediments in the understanding of the mechanisms of heat transfer with nanofluids is that, often, the experimental studies have been incomplete. Particle configuration/structure within the heterogeneous mixture is always absent in the measurements of viscosity and thermal conductivity. Particle sizes are not always measured during the measurements of the transport properties, even though several authors admit that particle aggregation may have influenced their data and conclusions. Most of the studies on nanofluids have neglected our previous experience and the vast amount of literature on gas–solids suspensions and solid–liquid fibers. This practice has resulted in some authors "reinventing the wheel." It also deprived some of the earlier experimental studies with a sound analytical basis and well-formulated mechanisms that would support the experimental data and would point to better and more systematic experimentation.

All the recent experimental and analytical evidence refutes the earlier stipulation that the addition of a small amount of nanoparticles causes an "anomalous" heat transfer enhancement. The heat transfer enhancement in nanofluids has been explained by the known mechanisms of heat transfer with particles and the existing theory for the conductivity of heterogeneous mixtures. It is rather unfortunate that the early experimental data on the subject were not analyzed with the then available analytical tools on conductivity and heat transfer theory. To paraphrase the late Carl Sagan, "Extraordinary claims require extraordinary evidence," and such evidence was neither given nor sought in many of the earlier studies of the past decade. The following are a few suggestions for the systematization and better coordination of future research on the subject, which will enable us to fully understand the heat transfer process with nanofluids:

- Use the theory of equilibrium thermodynamics for the rigorous definition of the properties of the fluid–solids mixtures and get rid of the ad hoc "models."
- Establish protocols for the preparation and preservation of the fluid–solid mixtures.
- Establish protocols for the measurement of the transport properties of the fluid–solids mixture, based on our expertise on the measurement of these properties for homogeneous mixtures and our knowledge of particulate behavior and flow.
- Recognize that the distribution/structure of the particles in the fluid matrix is one of the parameters that must be measured or inferred and always reported with the experimental data.

- Establish protocols for the simultaneous measurements of particle sizes and any other properties. Aggregation and flocculation processes continuously change the particle sizes in a mixture, and these processes affect the properties.
- Systematically question experimental data that do not agree with fundamental principles of heat transfer. For example, in convection Nu increases monotonically with Re, and the addition of highly conducting solids increases the conductivity of a mixture. Some experiments may need to be refined, reinterpreted, or repeated.

Bibliography

Altan CL, Elkatmis A, Yuksel M, Asian N, Bucak S (2011) Enhancement of thermal conductivity upon application of magnetic field to Fe_3O_4 nanofluids. J Appl Phys 110:093917-1–093917-7

Anoop KB, Kabelac S, Sundarajan T, Das SK (2009) Rheological and flow characteristics of nanofluids. J Appl Phys 106:034909

Bang IC, Chang SH (2005) Boiling heat transfer performance and phenomena of Al_2O_3-water nanofluids form a plain surface in a pool. Int J Heat Mass Transf 48:2407–2419

Basset AB (1888) Treatise on hydrodynamics. Bell, London

Batchelor G (1977) The effect of Brownian motion on the bulk stress in a suspension of spherical particles, I. Fluid Mech 83:97–117

Berg JC (2010) An introduction to interfaces and colloids - the bridge to nanoscience. World Scientific, Hackensack, NJ

Bergman TL (2009) Effect of reduced specific heats of nanofluids on single phase, laminar internal forced convection. Int J Heat Mass Transf 52:1240–1244

Bonnecaze RT, Brady JF (1990) A method for determining the effective conductivity of dispersions of particles. Proc Math Phys Sci 430(1879):285–313

Bonnecaze RT, Brady JF (1991) The effective conductivity of random suspensions of spherical particles. Proc Math Phys Sci 423(186):445–465

Brinkman H (1952) The viscosity of concentrated suspensions in solutions. J Chem Phys 20:571–582

Brock JR (1962) On the theory of thermal forces acting on aerosol particles. J Colloid Sci 17:768–780

Bruggeman DAG (1935) Berehnung vershidener physikalisher Konstanten von heterogenen Substanzen: I. Anallen der Physik 24:636–664

Brunn PO (1982) Heat or mass transfer from single spheres in a low Reynolds number flow. Int J Eng Sci 20(7):817–822

Buongiorno J, Venerus DC, Prabhat N, McKrell T, Townsend J, Christianson R, Tolmachev YV, Keblinski P, Hu L, Alvarado JL, Bang IC, Bishnoi SW, Bonetti M, Botz F, Cecere A, Chang Y, Chen G, Chen H, Chung SJ, Chyu MK, Das SK, Di Paola R, Ding Y, Dubois F, Dzido G, Eapen J, Escher W, Funfschilling D, Galand Q, Gao J, Gharagozloo PE, Goodson KE, Gutierrez JG, Hong H, Horton M, Hwang KS, Iorio CS, Jang SP, Jarzebski AB, Jiang Y, Jin L, Kabelac S, Kamath A, Kedzierski MA, Kieng LG, Kim C, Kim JH, Kim S, Lee SH, Leong KC, Manna I, Michel B, Ni R, Patel HE, Philip J, Poulikakos D, Reynaud C, Savino R, Singh PK, Song P, Sundararajan T, Timofeeva E, Tritcak T, Turanov AN, Vaerenbergh SV, Wen D, Witharana S, Yang C, Yeh WH, Zhao XZ, Zhou SQ (2009) A benchmark study on the thermal conductivity of nanofluids. J Appl Phys 106:094312

Chaudhri A, Lukes JR (2009) Multicomponent energy conserving dissipative particle dynamics: a general framework for mesoscopic heat transfer applications. J Heat Transf 131:033108-1–033108-9

Chen H, Ding Y, He Y, Tan C (2007) Rheological behavior of ethylene glycol based titania nanofluids. Chem Phys Lett 444:333–337

Cherkasova AS, Shan JW (2008) Particle aspect-ratio effects on the thermal conductivity of micro- and nanoparticle suspensions. J Heat Transf 130:082406-1–082406-7

Cherkasova AS, Shan JW (2010) Particle aspect-ratio and agglomeration-state effects on the effective thermal conductivity of aqueous suspensions of multiwalled carbon nanotubes. J Heat Transf 132:082402-1–082402-11

Choi SUS (2009) Nanofluids: from vision to reality through research. J Heat Transf 131:033106-1–033106-9

Choi SUS, Zhang ZG, Yu W, Lockwood FE, Grulke EA (2001) Anomalous thermal conductivity enhancement in nanotube suspensions. Appl Phys Lett 79:2252–2254

Choi CH, Ulmanella U, Kim J (2006) Effective slip and friction reduction in nanograted superhydrophobic microchannels. Phys Fluids 18:087105-1–087105-8

Chopkar M, Sudarshan S, Das PK, Manna I (2008) Effect of particle size on thermal conductivity of nanofluids. Metallurg Mater Trans A 39(7):1535–1542

Dansinger WJ (1963) Heat transfer to fluidized gas-solids mixtures in vertical transport. Ind Eng Chem Fundam 2(4):269–276

Das SK, Putra SK, Roetzel W (2003) Pool boiling characteristics of nanofluids. Int J Heat Mass Transf 46:851–862

Di Felice R, Rotondi M (2011) Fluid-particle drag force in binary-solid suspensions. Ind Eng Chem Res 011-01997f:1–37

Ding Y, Alias H, Wen D, Williams RA (2006) Heat transfer of aqueous suspensions of carbon nanotubes (CNT nanofluids). Int J Heat Mass Transf 49:240–250

Duan Z, Muzychka YS (2008) Slip flow heat transfer in annular microchannels with constant heat flux. J Heat Transf 130:092401-1–092401-8

Duan Z, Muzychka YS (2010) Effects of axial corrugated roughness on low Reynolds number slip flow and continuum flow in microtubes. J Heat Transf 132:041001-1–041001-9

Eapen J, Rusconi R, Piazza R, Yip S (2010) The classical nature of thermal conduction in nanofluids. J Heat Transf 132:102402-1–102402-14

Ebrahimnia-Bajestan E, Niazmand H, Duangthongsuk W, Wongwises S (2011) Numerical investigation of effective parameters in convective heat transfer of nanofluids flowing under a laminar flow regime. Int J Heat Mass Transf 54:4376–4388

EI-Genk MS, Yang IH (2008) Friction numbers and viscous dissipation heating for laminar flows of water in microtubes. J Heat Transf 130:082405-1–082405-13

Einstein A (1906) Eine neue Bestimmung der Molekuldimensionen. Ann Phys 19:289–306

Esparza HE (2012) Heat transfer enhancement in laminar microchannel flow by Monte Carlo simulations. M.S. Thesis, San Antonio, UT

Evans W, Fish J, Keblinski P (2006) Role of Brownian motion hydrodynamics on nanofluid thermal conductivity. Appl Phys Lett 88:093116

Fan J, Wang L (2011) Heat conduction in nanofluids: structure-property correlation. Int J Heat Mass Transf 54:4349–4359

Fantoni R, Giacometti A, Sciortino F, Pastore G (2010) Cluster theory of Janus particles. Soft Matter R Soc Chem 7:2419–2427

Farbar L, Depew CA (1963) Heat transfer effects to gas-solids mixtures in a circular tube. Ind Eng Chem Fundam 2:130–135

Frenkel N, Acrivos A (1967) On the viscosity of a concentrated suspension of solid spheres. Chem Eng Sci 6:847–853

Gao JW, Zheng R, Ohtani H, Zhu DS, Chen G (2009) Experimental investigation of heat conduction in nanofluids. Clue on clustering. Nano Lett 9:4128–4132

Ghadimi A, Saidur R, Metselaar HSC (2011) A review of nanofluid stability properties and characterization in stationary conditions. Int J Heat Mass Transf 54:4051–4068

Gibbs JW (1878) On the equilibrium of heterogeneous substances. In: The collective works of J. Willard Gibbs. Longmans, New York

Goicochea JV, Madrid M, Amon C (2010) Thermal properties for bulk silicon based on the determination of relaxation times using molecular dynamics. J Heat Transf 132:012401-1–012401-11

Granger R, Penninck S, Michaelides EE (2012) A critical review of nanofluid heat transfer – comparisons with particulate heat transfer on the macro-scale. In: ASME-IMECE, Houston, Nov 2012

Gupta SS, Siva VM, Krishnan S, Sreeprasad TS, Singh PK, Pradeep T, Das SK (2011) Thermal conductivity enhancement of nanofluids containing graphene nanosheets. J Appl Phys 110:084302-1–084302-6

Gurrum SP, King WP, Joshi YK, Ramakrishna K (2008) Size effect on the thermal conductivity of thin metallic films investigated by scanning joule expansion microscopy. J Heat Transf 130:082403-1–082403-8

Hamilton RL, Crosser OK (1962) Thermal conductivity of heterogeneous component systems. Ind Eng Chem Fundam 1:187–191

Hasofer AM, Lind NC (1974) Exact and invariant second moment code format. J Eng Mech Div ASCE 100:111–121

Heris SZ, Etemad SG, Esfahany MN (2006) Experimental investigation of oxide nanofluids laminar flow convective heat transfer. Int Commun Heat Mass Transf 33:529–535

Heyhat MM, Kowsary F (2010) Effect of particle migration on flow and convective heat transfer of nanofluids flowing through a circular pipe. J Heat Transf 132:062401-1–062401-9

Hong SW, Kang YT, Kleinstreuer C, Koo J (2011) Impact analysis of natural convection on thermal conductivity measurements of nanofluids using the transient hot-wire method. Int J Heat Mass Transf 54:3448–3456

Hosseini MS, Mohebbi A, Ghader S (2011) Prediction of thermal conductivity and convective heat transfer coefficient of nanofluids by local composition theory. J Heat Transf 133:052401-1–052401-9

Huang KH, Lee C-W, Wang C-K (2011) Boiling enhancement by TiO_2 nanoparticle deposition. Int J Heat Mass Transf 54:4895–4903

Jiji L (2008) Effect of rarefaction, dissipation, and accommodation coefficients on heat transfer in microcylindrical Couette flow. J Heat Transf 130:042404-1–042404-8

Ju YS, Kim J, Hung MT (2008) Experimental study of heat conduction in aqueous suspensions of aluminum oxide nanoparticles. J Heat Transf 130:092403-1–092403-6

Jung JY, Yoo JY (2009) Thermal conductivity enhancement of nanofluids in conjunction with electrical double layer (EDL). Int J Heat Mass Transf 52:525–528

Jung JY, Cho C, Lee WH, Kang YT (2011) Thermal conductivity measurement and characterization of binary nanofluids. Int J Heat Mass Transf 54:1728–1733

Kakaç S, Pramuanjaroenkij A (2009) Review of convective heat transfer enhancement with nanofluids. Int J Heat Mass Transf 52:3187–3196

Keblinski P, Phillpot SR, Choi SUS, Eastman JA (2002) Mechanisms of heat flow in suspensions of nano-sized particles (nanofluids). Int J Heat Mass Transf 45:855–863

Kestin J, Paul R, Shankland IR, Khalifa HE (1980) A high-temperature, high-pressure oscillating-disk viscometer for concentrated ionic solutions. Berichte der Bunsengesellschaft für physikalische Chemie 84:1255–1260. doi:10.1002/bbpc.19800841212

Khanafer K, Vafai K (2011) A critical synthesis of thermophysical characteristics of nanofluids. Int J Heat Mass Transf 54:4410–4428

Khiabani RH, Joshi Y, Aidun CK (2010) Heat transfer in microchannels with suspended solid particles: Lattice-Boltzmann based computations. J Heat Transf 132:041003-1–041003-9

Kim J-K, Jung JY, Kang YT (2006) The effect of nano-particles on the bubble absorption performance in a binary nanofluid. Int J Refrigeration 29:22–29

Kim S, Kim C, Lee WH, Park SR (2011) Rheological properties of alumina nanofluids and their implication to the heat transfer enhancement mechanism. J Appl Phys 110:034316-1–034316-6

Kolade B, Goodson KE, Eaton JK (2009) Convective performance of nanofluids in a laminar thermally developing tube flow. J Heat Transf 131:052402-1–052402-8

Komati S, Suresh AK (2009) Anomalous enhancement of interphase transport rates by nanoparticles: effect of magnetic iron oxide on gas-liquid mass transfer. Ind Eng Chem Res 49(1):390–405

Kondaraju S, Jin EK, Lee JS (2010) Direct numerical simulation of thermal conductivity of nanofluids: the effect of temperature two-way coupling and coagulation of particles. Int J Heat Mass Transf 53:862–869

Koo J, Kleinstreuter C (2004) A new thermal conductivity model for nanofluids. J Nanopart Res 6:577–588

Koo J, Kleinstreuter C (2005a) Impact analysis of nanoparticle motion mechanisms on the thermal conductivity of nanofluids. Int Commun Heat Mass Transf 32:1111–1118

Koo J, Kleinstreuter C (2005b) Laminar nanofluid flow in microheat sinks. Int J Heat Mass Transf 48:2652–2661

Krieger IM (1959) A mechanism for non-newtonian flow in a suspension of rigid spheres. Trans Soc Reol 3:137–152

Krieger IM (1972) Rheology of polydisperse latices. Adv Colloid Interface Sci 71:622–624

Krieger IM, Eguiluz M (1976) The second electroviscous effect in polymer lattices. Trans Soc Rheol 20:29–45

Krishna K, Ganapathy H, Sateesh G, Das S (2011) Pool boiling characteristics of metallic nanofluids. J Heat Transf 133:111501-1–111501-8

Kumar DH, Patel HE, Kumar VRR, Sundararajan T, Pradeep T, Das SK (2004) Model for heat conduction in nanofluids. Phys Rev Lett 93:144301-1–144301-4

Kuravi S, Kota KM, Du J, Chow LC (2009) Numerical investigation of flow and heat transfer performance of nano-encapsulated phase change material slurry in microchannels. J Heat Transf 131:062901-1–062901-9

Lai WY, Vinod S, Phelan PE, Prasher R (2009) Convective heat transfer for water-based alumina nanofluids in a single 1.02-mm tube. J Heat Transf 131:112401-1–112401-9

Lance M, Bataille J (1991) Turbulence in the liquid phase of a uniform bubbly air-water flow. J Fluid Mech 222:95–118

Lee D, Kim JW, Kim BG (2006) A new parameter to control heat transport in nanofluids: surface charge state of the particle in suspension. J Phys Chem B 110:4323–4328

Lee HJ, Liu DY, Alyousef Y, Yao S (2010a) Generalized two-phase pressure drop and heat transfer correlations in evaporative micro/minichannels. J Heat Transf 132:041004-1–041004-9

Lee J, Gharagozloo P, Kolade B, Eaton J, Goodson K (2010b) Nanofluid convection in microtubes. J Heat Transf 132:092401-1–092401-5

Lee SW, Park SD, Kang S, Bang IC, Kim JH (2011) Investigation of viscosity and thermal conductivity of SiC nanofluids for heat transfer applications. Int J Heat Mass Transf 54:433–438

Li T, Benyahia S (2011) Revisiting Johnson and Jackson boundary conditions for granular flows. AIChE J 57:1–11

Li J, Kleinstreuter C (2010) Entropy generation analysis for nanofluid flow in microchannels. J Heat Transf 132:122401-1–122401-8

Li CH, Peterson GP (2007) Mixing effect on the enhancement of the effective thermal conductivity of nanoparticle suspensions (nanofluids). Int J Heat Mass Transf 50:4668–4677

Li X, Wang T (2008) Two-phase flow simulation of mist film cooling on turbine blades with conjugate internal cooling. J Heat Transf 130:102901-1–102901-8

Liu MS, Lin MCC, Tsai CY, Wang CC (2006) Enhancement of thermal conductivity with Cu for nanofluids using chemical reduction method. Int J Heat Mass Transf 49:3028–3033

Lohse D, Luther S, Rensen J, van den Berg TH, Mazzitelli I, Toschi F (2004) Turbulent bubbly flow. In: Proceedings of 5th international conference on multiphase flow, Yokohama, Japan

Lundgren TS (1972) Slow flow through stationary random beds and suspensions of spheres. J Fluid Mech 51:273–299

Luo T, Lloyd JR (2008) Ab initio molecular dynamics study of nanoscale thermal energy transport. J Heat Transf 130:122403-1–122403-7

Mahbubul IM, Saidur R, Amalina MA (2012) Latest developments on the viscosity of nanofluids. Int J Heat Mass Transf 55:874–885

Masoumi N, Sohrabi N, Behzadmehr A (2009) A new model for calculating the effective viscosity of nanofluids. J Appl Phys 42(055501):1–6

Maxwell JC (1881) A treatise on electricity and magnetism, 2nd edn. Clarendon, Oxford

Mazzitelli I, Venturoli M, Melchionna S, Succi S (2011) Towards a mesoscopic model of water-like fluids with hydrodynamic interactions. J Chem Phys 135:124902-1–124902-10

Michaelides EE (1986) Heat transfer in particulate flows. Int J Heat Mass Transf 29(2):265–273

Michaelides EE (2006) Particles, bubbles and drops – their motion, heat and mass transfer. World Scientific, Hackensack, NJ

Michaelides EE, Feng Z-G (1994) Heat transfer from a rigid sphere in a non-uniform flow and temperature field. Int J Heat Mass Transf 37:2069–2076

Millikan RA (1923) The general law of fall of a small spherical body through a gas and its bearing upon the nature of molecular reflection from surfaces, Phys. Phys Rev 22:1–23

Mooney M (1951) The viscosity of a concentrated suspension of spherical particles. J Colloid Sci 6:162–170

Nan CW, Birringer R, Clarke DR, Gleiter H (1997) Effective thermal conductivity of particulate composites with interfacial thermal resistance. J Appl Phys 81:6692–6699

Nan CW, Shi Z, Lin Y (2003) A simple model for thermal conductivity of carbon nanotube-based composites. Chem Phys Lett 375:666–669

Nedea SV, Markvoort AJ, van Steenhoven AA, Hilbers PAJ (2009) Heat transfer predictions for micro/nanochannels at the atomistic level using combined molecular dynamics and Monte Carlo techniques. J Heat Transf 131:033104-1–033104-8

Nguyen CT, Desgranges F, Roy G, Galanis N, Marie T, Boucher S, Mintsa HA (2007) Temperature and particle-size dependent viscosity data for water based nanofluids–hysteresis phenomenon. Int J Heat Fluid Flow 28:1492–1506

Nie C, Marlow WH, Hassan YA (2008) Discussion of proposed mechanisms of thermal conductivity enhancement in nanofluids. Int J Heat Mass Transf 51:1342–1348

O'Hanley H, Buongiorno J, McKrell T, Ho L (2012) Measurement and model validation of nanofluid specific heat capacity with differential scanning calorimetry. Adv Mech Eng. doi:10.1155/2012/181079

Olle B et al (2006) Enhancement of oxygen mass transfer using functionalized magnetic nanoparticles. Ind Eng Chem Res 45(12):4355–4363

Pak B-C, Cho Y (1998) Hydrodynamic and heat transfer study of dispersed fluids with submicron metallic oxide particles. Exp Heat Transf 11:151–170

Pfeffer R, Rosetti S, Liclein S (1966) Analysis and correlation of heat transfer coefficient and friction factor data for dilute gas-solids suspensions. NASA report TN D-3603. NASA, Washington, DC

Philip J, Sharma PD, Raj B (2008) Evidence for enhanced thermal conduction through percolating structures in nanofluids. Nanotechnology 19:305706

Prasher R, Bhattacharya P, Phelan PE (2005) Thermal conductivity of nanoscale colloidal solutions (nanofluids). Phys Rev Lett 94:025901

Prasher RS, Bhattacharya P, Phelan PE (2006a) Brownian motion based convective-conductive model for the effective thermal conductivity of nanofluids. J Heat Transf 128:588–595

Prasher R, Evans W, Meaking P, Fish J, Phelan P, Keblinski P (2006b) Effects of aggregation on thermal conduction in colloidal nanofluids. Appl Phys Lett 89:143119

Probstein RF (1994) Physicochemical hydrodynamics, 2nd edn. Elsevier, New York

Randrianalisoa J, Baillis D (2008) Monte Carlo simulation of steady-state microscale phonon heat transport. J Heat Transf 130:072404-1–072404-13

Russel WR, Saville DA, Schowalter WR (1989) Colloidal dispersions. Cambridge University Press, Cambridge

Sarkar S, Selvam RP (2009) Direct numerical simulation of heat transfer in spray cooling through 3D multiphase flow modeling using parallel computing. J Heat Transf 131:121007-1–121007-8

Schluderberg DC, Whitelaw RL, Carlson RW (1961) Gaseous suspensions – a new reactor coolant. Nucleonics 19:67–76

Shin D, Banerjee D (2010) Effects of silica nanoparticles on enhancing the specific heat capacity of carbonate salt eutectic. Int J Struct Changes Solids 2:25–31

Shin D, Banerjee D (2011) Enhancement of specific heat capacity of high-temperature silica-nanofluids synthesized in alkali chloride salt eutectics for solar thermal-energy storage applications. Int J Heat Mass Transf 54:1064–1070

Shukla R, Dhir V (2008) Effect of Brownian motion on thermal conductivity of nanofluids. J Heat Transf 130:042406-1–042406-13

Talbot L, Cheng RK, Schefer RW, Willis DR (1980) Thermophoresis of particles in a heated boundary layer. J Fluid Mech 101:737–758

Taylor TD (1963) Heat transfer from single spheres in a low Reynolds number slip flow. Phys Fluids 6(7):987–992

Tien CL, Lienhard JH (1979) Statistical thermodynamics, revised. Hemisphere, New York

Timofeeva EV, Gavrilov AN, McCloskey JM, Tolmachev WV, Sprunt S, Lopatina LM, Sellinger JV (2007) Thermal conductivity and particle agglomeration in alumina nanofluids: experiment and theory. Phys Rev E 76:061203

Torii S, Yang WJ (2009) Heat transfer augmentation of aqueous suspensions of nano-diamonds in turbulent pipe flow. J Heat Transf 131:043203-1–043203-5

Torii D, Ohara T, Ishida K (2010) Molecular-scale mechanism of thermal resistance at the solid-liquid interfaces: influence of interaction parameters between solid and liquid molecules. J Heat Transf 132:012402-1–012402-9

Tran-Cong S, Gay M, Michaelides EE (2004) Drag coefficients of irregularly shaped particles. Powder Technol 139:21–32

Tseng WJ, Chen CN (2003) Effect of polymeric dispersant on rheological behavior of nickel–terpineol suspensions. Mater Sci Eng A347:145–153

Tseng WJ, Lin KC (2003) Rheology and colloidal structure of aqueous TiO_2 nanoparticle suspensions. Mater Sci Eng A355:186–192

Tzou DY (2008) Instability of nanofluids in natural convection. J Heat Transf 130:072401-1–072401-9

Vrabec J, Horsch M, Hasse H (2009) Molecular dynamics based analysis of nucleation and surface energy of droplets in supersaturated vapors of methane and ethane. J Heat Transf 131:043202-1–043202-4

Wachtell GP, Wagener JP, Steigelman WH (1961) Evaluation of gas-graphite suspensions as nuclear reactor coolants. Report NYO-9672 AEC. Franklin Institute, Philadelphia, PA

Wamkam CT, Opoku MK, Hong H, Smith P (2011) Effects of pH on heat transfer nanofluids containing ZrO_2 and TiO_2 nanoparticles. J Appl Phys 109:024305-1–024305-5

Wang L, Wei X (2009) Nanofluids: synthesis, heat conduction, and extension. J Heat Transf 131:033102-1–033102-7

Wang X, Xu X, Choi SUS (1999) Thermal conductivity of nanoparticle fluid mixture. J Thermophys Heat Transf 13:474–480

Wen D (2008) Mechanisms of thermal nanofluids on enhanced critical heat flux (CHF). Int J Heat Mass Transf 51:4958–4965

Wen D, Ding Y (2004) Experimental investigation into convective heat transfer of nanofluids at the entrance region under laminar flow conditions. Int J Heat Mass Transf 47:5181–5188

Wen D, Ding Y (2005) Experimental investigation into the pool boiling heat transfer applications. Int J Heat Fluid Flow 26:855–864

Williams W, Buongiorno J, Hu LW (2008) Experimental investigation of turbulent convective heat transfer and pressure loss of alumina/water and zirconia/water nanoparticle colloids (nanofluids) in horizontal tubes. J Heat Transf 130:042412

Xuan Y, Li Q (2003) Investigation on convective heat transfer and flow features of nanofluids. J Heat Transf 125:151–155

Xue QZ (2003) Model for effective thermal conductivity of nanofluids. Phys Lett A 307:313–317

Xue QZ, Keblinski P, Philpot SR, Choi SUS, Eastman JA (2004) Effect of liquid layering at the liquid–solid interface on thermal transport. Int J Heat Mass Transf 47:4277–4284

Yang B (2008) Thermal conductivity equations based on Brownian motion in suspensions of nanoparticles (Nanofluids). J Heat Transf 130:042408-1–042408-5

You SM, Kim JH, Kim KH (2003) Effect of nanoparticles on critical heat flux of water in pool boiling heat transfer. Appl Phys Lett 83:3374–3376

Index

E.E. (Stathis) Michaelides, *Heat and Mass Transfer
in Particulate Suspensions*, SpringerBriefs in Applied Sciences and Technology,
DOI 10.1007/978-1-4614-5854-8, © Springer Science+Business Media New York 2013